ENVIRONMENT AND SOCIETY

Environment and Society
Sustainability, Policy and the Citizen

STEWART BARR
University of Exeter, UK

Published by
Ashgate Publishing Limited
Gower House
Croft Road
Aldershot
Hampshire GU11 3HR
England

Ashgate Publishing Company
Suite 420
101 Cherry Street
Burlington, VT 05401-4405
USA

Ashgate website: http://www.ashgate.com

British Library Cataloguing in Publication Data
Barr, Stewart, 1976-
    Environment and society : sustainability, policy and the
    citizen. - (Ashgate studies in environmental policy and
    practice)
    1. Environmentalism - Social aspects 2. Environmental
    protection - Citizen participation 3. Environmental
    responsibility
    I. Title
    304.2

Library of Congress Cataloging-in-Publication Data
Barr, Stewart, 1976-
    Environment and society : sustainability, policy and the citizen / by Stewart Barr.
        p. cm.
    Includes bibliographical references and index.
    ISBN: 978-0-7546-4343-2
    1. Human ecology. 2. Social ecology. 3. Environmentalism--Social aspects. 4.
    Sustainable development. I. Title.

    GF41.B35 2008
    304.2--dc22

                                                              2007035684

ISBN: 978-0-7546-4343-2

Printed and bound in Great Britain by MPG Books Ltd, Bodmin, Cornwall.

Contents

List of Figures   *vii*
List of Tables   *ix*
Foreword   *xi*
Preface   *xiii*
Acknowledgements   *xv*

PART 1: CONTEXTS

1   Green Dilemmas   3

2   Sustainability   21

3   Policy   51

PART 2: PERSPECTIVES

4   Behaviour Change: Policy and Practice   85

5   The Social Psychology of Environmental Action   105

PART 3: APPROACHES

6   Framing Environmental Practice   129

7   Sustainable Lifestyles   161

8   The Value-Action Gap   191

PART 4: APPLICATIONS

9   Changing Behaviour: A Social Marketing Approach   229

10   Sustainability, Citizens and Progress   251

Bibliography   *263*

Index   *279*

List of Figures

1.1   *Human perspectives* (after Meadows et al. 1972)   5

1.2   The Malthusian cycle of population growth and decline   12

1.3   'World model standard run' from *The Limits to Growth* (after Meadows et al. (1972)   14

1.4   Scenario from *The Limits to Growth* assuming a doubling of natural resource reserves (after Meadows et al. 1972)   15

2.1   Key concepts for exploring sustainable development   32

2.2   Khan's (1995) paradigm of sustainable development   39

2.3   The 'three rings' model of sustainability (after Giddings et al. 2002)   41

2.4   Giddings et al.'s (2002) nested sustainability model   43

2.5   Weak and strong approaches to sustainable development (after Roberts 2004)   44

2.6   The sustainability spectrum (after Pearce 1993)   46

3.1   Policy structures for sustainable development   53

3.2   Conflicts and sustainability   55

3.3   Headline indicators of sustainable development between 1999 and 2004 (DETR, 1999c)   62

3.4   Guiding principles from *Securing the Future* (adapted from DEFRA 2005)   67

3.5   A framework for behaviour change presented in *Securing the Future* (adapted from DEFRA 2005)   68

3.6   The 20 framework indicators for sustainable development   70

3.7   National, regional and local delivery structures for sustainable development in the UK (adapted from DEFRA 2005)   72

3.8   Sustainable development: the planning policy framework   73

3.9   Local Agenda 21 logo from Exeter City Council's LA21 Strategy (adapted from Exeter City Council 1996)   79

4.1   Community Action 2020 (adapted from DEFRA 2005)   91

4.2   The Theory of Reasoned Action (based on Fishbein and Ajzen, 1975)   101

4.3   A conceptual framework of environmental behaviour   102

6.1   Study locations in Devon, South West England   133

6.2   Energy-saving behaviour in the survey   145

6.3   Energy-efficiency behaviour in the survey   145

6.4   Water conservation behaviour in the survey   147

6.5   Use of water saving devices in the survey   147

6.6   Green consumption behaviours in the survey   149

6.7   Waste management behaviours in the survey   151

6.8   Analytical structure of the book   159

7.1   Reported behaviour for the Committed Environmentalists cluster   166

7.2   Reported behaviour for the Mainstream Environmentalists cluster   167
7.3   Reported behaviour for the Occasional Environmentalists cluster   168
7.4   Reported behaviour for the Non-environmentalists cluster   169
8.1   Path analysis of purchase related behaviour   198
8.2   Path analysis of habitual behaviour   199
8.3   Path analysis of recycling behaviour   200
8.4   Path analysis of purchase related behaviour for cluster 1
       (Committed Environmentalists)   205
8.5   Path analysis of purchase related behaviour for cluster 2
       (Mainstream Environmentalists)   206
8.6   Path analysis of purchase related behaviour for cluster 3
       (Occasional Environmentalists)   207
8.7   Path analysis of purchase related behaviour for cluster 4
       (Non-environmentalists)   208
8.8   Path analysis for habitual behaviour cluster 1
       (Committed Environmentalists)   211
8.9   Path analysis for habitual behaviour cluster 2
       (Mainstream Environmentalists)   212
8.10   Path analysis for habitual behaviour for cluster 3
       (Occasional Environmentalists)   213
8.11   Path analysis for habitual behaviour for cluster 4
       (Non-environmentalists)   214
8.12   Path analysis for recycling behaviour cluster 1
       (Committed Environmentalists)   217
8.13   Path analysis for recycling behaviour cluster 2
       (Mainstream Environmentalists)   218
8.14   Path analysis for recycling behaviour for cluster 3
       (Occasional Environmentalists)   219
8.15   Path analysis for recycling behaviour for cluster 4
       (Non-environmentalists)   220

List of Tables

6.1	Reported behaviours measured in the survey	136
6.2	Behavioural intention items measured in the survey	138
6.3	Social values measured in the survey	139
6.4	Environmental values measured in the survey	140
6.5	Environmental knowledge quiz in the survey	141
6.6	Sources of information about environmental issues in the survey	142
6.7	Behavioural factors derived from Principal Component Analysis	154
7.1	Demographic composition of the sample by cluster membership	172
7.2	Factorially-defined social values of the sample by cluster membership	177
7.3	Factorially-defined environmental values in the sample by cluster membership	178
7.4	Factorially-defined psychological variables in the sample by cluster membership	180
8.1	Multiple regression models for behaviour and behavioural intention according to the three types of environmental practices	196
8.2	Multiple regression models for behaviour and behavioural intention according to the three types of environmental practices and four behavioural clusters	203

Foreword

This book seeks to explore one of the most pressing and complex questions of the early twenty-first century: how to promote the behavioural shifts necessary for creating the 'sustainable society'? Through a critical approach to the links between sustainability, policy and citizen engagement, the book argues that sustainability policy needs to undergo a major conceptual shift, moving away from a negative approach to behaviour change towards a positive perspective, utilising the well-known techniques of segmentation and social marketing. Such 'mainstreaming' of sustainable lifestyles is likely to be the only effective means of engaging the majority of citizens in the environmental debate, given the major influence of the consumer society on individual aspirations and beliefs.

The book begins by setting out three 'green dilemmas' concerned with the conflicts between conservation and growth, the 'ruptured' scales of global change and personal activism and the tension between the 'needs' of individuals and society. Using these three dilemmas as framing contexts, the book explores how the significance of 'the individual' to sustainability debates has increased both in the conceptual and political development of sustainability policy. Conceptually, this has witnessed the importance of 'bottom-up' approaches to resolving environmental dilemmas, whilst politically there has been recognition that individual citizens hold the key to meeting critical environmental targets through changes in their lifestyles.

The book then moves on to explore the academic basis for examining 'environmental action', a contested set of behavioural responses from recycling household materials to 'green purchasing' which have been promoted by policy makers as the most effective way for citizens to help the environment. This entails a critical re-appraisal of current research within both policy and geography for encouraging behaviour change and argues that the significant methodological and epistemological differences between these approaches needs to be overcome. Accordingly, the book calls for a closer engagement by geographers and policy makers with existing research from social-psychology that has explored the links between environmental attitudes and behaviour using primarily quantitative methods and from a more holistic perspective. On the basis of this approach, a conceptual framework of environmental behaviour is outlined and through the use of data collected from two major research projects, a new perspective on exploring and promoting sustainable lifestyles is discussed.

Chapters 6 to 9 demonstrate the utility of taking a more holistic perspective towards framing, segmenting, explaining and prompting environmental action through the use of data collected during these two major research projects from 2001 to 2006. The research provides evidence that environmental practices are intimately related to everyday practices in and around the home and that a useful way of exploring such differences is through the use of examining 'lifestyle' groups, which demonstrate specific behavioural and attitudinal characteristics. Such groups can be used to examine the influence of specific environmental practices more effectively and, through the utilisation of a social marketing approach, change the behaviour of particular lifestyle groups.

Accordingly, the book provides a new intellectual and practical agenda for exploring personal commitments towards the environment, emphasising the importance of relating environmental practice to everyday practice and the need to target policy more effectively towards lifestyle groups with specific characteristics. The book calls for a new research agenda to examine the role of social marketing in promoting environmental behaviour and the need to examine the longitudinal nature of behaviour change across a series of environmental practices.

Preface

In recent years and most notably in the months leading up to the publication of this book, environmental issues have featured prominently in the media. It seems that both the predictions of future climate change and their potential impacts have led to a greater focus on the contribution that individual citizens can make to mitigate against the effects of global climate change. The media focus is of course subject to constant change: in 2006 and 2007 the focus was on the 'problem' of using low cost airlines; we can only speculate what the next issue might be. However, whatever arises, it is certain that the role of individual citizens in determining our environmental future is firmly on the political agenda.

This state of affairs has emerged relatively quickly; during the research that formed the basis of this book (started some six years ago), the 'environmental action' debate mostly focussed on localised environmental issues such as recycling. We are now faced with a growing political stampede to promote the green credentials of political parties and this race to be green will of course only be tested when a general election is forthcoming – being 'green' is easy in the middle of a parliamentary term, less so when votes are at stake.

So what are the prospects for the new-found green agenda? One view is to argue that the effect of a general election will be to sideline the environmental agenda, as has always been the case – 'It's the economy, stupid' as one US Presidential candidate put it. Or have we progressed a little further than this? Have we indeed crossed a Rubicon that indicates a point of no return for the environmental agenda? It is probably the case that, for the first time in its history, the environmental movement has generated enough momentum to keep going for at least the medium term. In part, this is due to the media's focus on environmental issues and in particular global warming. However, it is also the result of a greater mainstreaming of the environmental movement, helped by the increasing levels of professionalism associated with green organisations.

This second point, 'mainstreaming' (an ugly, but currently popular term in Whitehall), is likely to be the key to success for those promoting an environmental agenda. If the government is really committed to changing peoples' lifestyles to make them more sustainable, then it will have to speak to more than a small segment of the population, already convinced on the benefits of living a greener lifestyle. It must speak to the majority. Yet speaking to the majority means speaking a different language and, critically, speaking to peoples' existing values and aspirations. We must move away from the messages that argue being green means having a lower standard of living or implies 'switching off the lights' for good. In the consumer society, individuals listen to the most compelling messages from sources that they trust. The environmental movement, business and the government need to develop messages to speak to the consumer society. Evidence already exists from the success of the 'Fair Trade' movement that this can be achieved. However, for business and government, this also means change. Mistrust in politicians and 'big business' is significant and these institutions will also need to demonstrate that they are sincere about changing their own practices.

The message is therefore a positive one. There are real opportunities for promoting behaviour change amongst the population. However, just as the social science of exploring the notion of behaviour change is uncertain, the science of environmental change and technological development is also rapidly developing and subject to change. So whilst 'behaviour change' is currently on the political agenda for the short-to-medium term, we must acknowledge that new developments and discoveries are likely to mould the arguments presented here in ways which (sitting in Exeter on a wet July day in 2007) none of us can predict. One thing does remain certain however: understanding human behaviour remains as challenging and complex as any study of the natural environment.

Stewart Barr
Exeter, July 2007

Acknowledgements

I would like to thank a number of former colleagues from the Department of Geography at Exeter for their help and support during the production of this book. In particular, I would like to pay tribute to Andrew Gilg and Nicholas Ford, who secured funding from the Economic and Social research Council in the form of a two-year research grant which forms the empirical basis for much of the book. I would also like to thank Andrew Gilg and Gareth Shaw for their support in attaining a DEFRA-funded research grant, material from which has been used to write Chapters 8 and 9 of this book. I am grateful to both of them for permission to use this material and in particular to Andrew who provided invaluable comments on an early draft of this book. Both Andrew and Gareth, although no longer in the Department of Geography, have continued to work with me on developing a new agenda for exploring sustainable lifestyles in a post-disciplinary context and I am delighted that Andrew continues his work as Professor of Rural Planning at Gloucester University and Gareth as Professor of Tourism and Retail Management at Exeter.

I would also like thank Sue Rouillard from the Department of Geography's Drawing Office. Her professionalism and responsiveness to my seemingly endless requests for new and altered diagrams are greatly appreciated. I have also appreciated the assistance from colleagues across the academy for their helpful contribution and permission to use copyright material.

There are also other colleagues, friends and family who are too numerous to mention by name who have given considerable support during the length of this project, which at times has seemed to be without end.

PART 1
Contexts

Chapter 1

Green Dilemmas

Environment and Economy

Not since the early 1970s, the period known colloquially as the Environmental Crisis, has the 'environment' featured so prominently in the popular news media and within mainstream political discourse. This rediscovery of the green agenda has been viewed by many as the rebirth of a distinct political discourse that reflects increased environmental concern amongst the population and a re-engagement with the environmental agenda by politicians and policy-makers. Domestically, the green agendas displayed by all three of the mainstream political parties in the United Kingdom have highlighted the need to tackle environmental degradation. Such a rapid shift in political attitudes towards the environment has been framed almost solely around the need to tackle climate change, a threat perceived as having not only environmental, but crucial social and economic impacts. To this end, the British Government has shifted emphasis away from making an environmental case for action to combat climate change towards an economic argument recently evidenced in the Stern Review on the economics of climate change (HM Treasury 2006). Politically, competition amongst party leaders has been fierce to present the most environmentally sound policies and all parties have highlighted key shifts that will be required to tackle global environmental change, including the use of regulation, financial penalties, incentives and exhortation (Gilg 2005) as means by which to effect change. For example, the Labour administration's focus on taxing specific 'environmental bads' has firmly grounded the principle of penalties in the mind of businesses and commerce. In other instances, policies have been focused around engaging the public in more environmentally sound activities, such as recycling or energy conservation. Yet critics have pointed to the obvious flaws in such strategies, not least the dilemma that lies at the heart of attempting to implement such policies which many have argued will be detrimental to economic progress.

This first 'green dilemma' for twenty-first century society can be illustrated with reference to the growing air travel industry within the United Kingdom. Since the late 1990s low cost airlines have emerged as a distinctive force in budget travel, offering frequent flights from an expanding range of regional airports to both domestic and European destinations. What has characterised the growth of such airlines is the vigorous marketing techniques and the use of Internet technology to promote low cost, short-haul air travel. This has witnessed the growth of a range of market segments, not least the major increase in short breaks and city breaks, alongside 'stag and hen' weekends, 'spa breaks', skiing, rugby, football and golfing pursuits. In many instances, the growth in short-haul air travel has either replaced existing modes of transport, such as rail or coach travel, or has seen the emergence of new markets. In any case, the increase in air travel has had a marked impact on passenger numbers. For example, at London Luton Airport, passenger figures more than quadrupled in the ten years 1994 to 2004 (DfT 2005), due in the main to the low cost airline easyjet. Indeed, the environmental impact of such increased air travel has been the subject of recent vexed debate. Statistics are regularly quoted regarding the amount of kilograms of carbon each passenger will emit for a specific journey. The very act of boarding an aeroplane for whatever duration is contextualised and deliberated according to the environmental damage that this specific act creates. This is presented at the broadest spatial scale and is rarely conceptualised within temporal constraints. The message is therefore a simple one: flying is 'bad' (a message now being countered by the first attempts of the airline industry to provide eco-labelling for customers).

Accordingly, the first 'green dilemma' we encounter in modern society is the desire for ongoing economic growth alongside an acknowledgement that such economic progress may be causing damage which could have a fundamental impact on the global environmental system. For some, the remedy to this problem is simple: economic reductionism and a move to frugality – even perhaps a move away from capitalism – as a means by which to safeguard future generations from environmental catastrophe. Yet this simplistic message, whilst providing a clear and concise direction for behavioural change, is likely to resonate with no more than a minority of citizens, already committed to the environment.

Human Perspectives: Time and Space

Meadows et al. (1972)[1] succinctly demonstrated why this was likely to be the case, by highlighting both the spatial and temporal constraints on human beings to conceptualise and act on environmental threats. Figure 1.1 is adapted from the original suggested by Meadows et al. (1972) and demonstrates how the human ability to perceive and act on threats at the 'family' scale and within a short timeframe are acutely sensed, whilst those at a global and generational scale are perceived very weakly. Meadows et al. (1972, 18) argued that such perceptions necessarily varied from person to person, stating that 'In general the larger the space and the longer the time associated with a problem, the smaller the number of people who are actually concerned about its situation'. Although Meadows et al. argued that such differences in perception would differ according to culture and experience, there is an implicit belief in western societies that the advanced stages of development achieved by many states enables them and their populations to consider global problems in a way that others simply cannot. However, it is becoming widely acknowledged in western societies that concerns expressed at

---

1   Please note that Meadows et al. have updated their 1972 text: Meadows, D.H. et al. (2004) *Limits to Growth – the 30-Year Update* (White River Jct, VT: Chelsea Green Publishing).

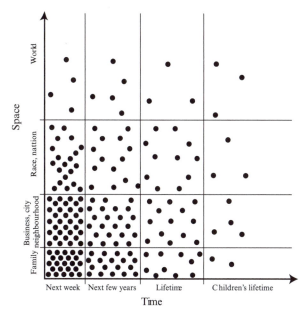

Figure 1.1 *Human perspectives* **(after Meadows et al. 1972). Reproduced with the kind permission of Professor Dennis Meadows**

Source: Meadows et al. (1972).

the global level and in terms of generational timeframes have little resonance with the everyday lives and practices of individuals.

This disconnection between lived experiences in time and space is the second 'green dilemma' and what forms the basis for the intellectual and practical challenges that face western societies attempting to grapple with the enormities of environmental issues and their potential amelioration. Goodwin and Barr (2007) have referred to these (multiple) disconnections as the 'rupturing of scale', reflected in and between scientific, political and lay discourses on the environment. Within the context of this book, a key dilemma which faces western societies is the 'rupture' between the lived experience of 'environment' or 'nature' and the global context. Eden (1993) has argued that as individuals we have become desensitised to environmental change through the developments in technology that have insulated us from the environment. Apart from the aesthetics of green space in our mainly urbanised living environments, we may only experience 'nature' when engaging in tourism and recreation. Even then, this may be a highly sanitised and artificial experience.

In temporal terms, we have become concerned more with our ability to consume and take ownership of resources for immediate gratification. The notion of expending effort and investment for long-term gain has been steadily replaced with a timescale that permits decisions to be taken only at a political timescale of perhaps four to five years. At the individual scale, this is reflected in the common decisions by householders not to invest in energy saving devices and consumables, such as more efficient boilers or light bulbs, because the immediate 'gain' is not evident. More widely, decisions to invest in environmental technology or more sustainable infrastructure are sidelined within the political context of the day.

Accordingly, the second green dilemma seeks to emphasise the critical problems of 'scale'; both the spatial scale of environmental problems, their cause and potential resolution, alongside the temporal scaling of environmental problems within and between generations. It is of course intimately related to the first dilemma (conservation versus growth) because decisions on the level of economic growth are framed at both spatial and temporal scales. The final green dilemma builds on the first two by exploring the role of individuals within the global system, illustrating the inherent tension between individual and societal interests, a topic which will become the main focus of this book.

A Tragedy for Our Commons? Individuals and Society

To explore this last dilemma we must turn to what Garrett Hardin (1968) institutionalised as the concept of the *Tragedy of the Commons,* a long-established notion of over-consumption which presents as much a dilemma for today's industrialised societies as it invoked for Hardin's specific example. Hardin explained a state of resource exploitation and the subsequent impacts with an emotive and effective account of pasture land and grazing. The 'tale' begins with what we can term resource equilibrium; each herdsman on the pasture keeps as many cattle on the common as is possible. The equilibrium between resource exploitation (in the form of grass grazed) is maintained due the nature of the society on which the use of the pasture is based. Tribal wars, famines, disease and poaching ensure that the number of herdsmen and cattle remain at a level at which resources can be renewed. However, this 'idyllic' (in resource terms) situation comes to an end 'Finally, however, comes the day of reckoning, that is, the day when the long-desired goal of social stability becomes a reality. At this point, the inherent logic of the commons remorselessly generates tragedy' (Hardin 1968, 1244). The 'social' stability that Hardin refers to has numerous parallels in contemporary society, and in particular is reflected in the development cycles of nations. Social stability can be represented in terms of both political and social goals. Politically, one can infer the development of stable regimes and democratic government. Socially, perhaps the greatest area of stability in the short term would refer to health, with increasing life expectancies and, in the short term, higher birth rates. In the medium and long terms, social stability is more likely to be a function of welfare, with an increasing emphasis on quality of life, economic prosperity and 'higher order needs' (Maslow 1970).

Returning to Hardin's tale, these changed social conditions alter the implicit instincts of the herdsman. Confronted with more desirable social conditions and stability, Hardin (1968, 1244) argues that 'As a rational being, each herdsman seeks to maximize his gain. Explicitly or implicitly, more or less consciously, he asks "What is the utility *to me* of adding one more animal to my herd?"' (original emphasis). Hardin makes clear that as a rational economic being, the herdsman's first instinct is to turn to the gain that he might personally attain from any change in his herd. The answer to this question is clear in Hardin's eyes. Positively, the herdsman has the benefit of one additional animal, which includes the sale of the animal. Negatively, the over-grazing of one animal is of little consequence, since the over-grazing will be shared by all the herdsmen, thus minimising any impact. Such a calculation inevitably leads the herdsman to consider adding additional animals to the common, since the negative impacts are minor compared to the positive effect of adding animals:

> But this is the conclusion reached by each and every rational herdsman sharing a commons. Therein is the tragedy. Each man is locked into a system that compels him to increase his herd without limit – in a world that is limited. Ruin is the destination towards which all men rush, each pursuing his own best interest in a society that believes in the freedom of the commons. Freedom in a commons brings ruin to all. (Hardin 1968, 1244)

Hardin's tale is a popular one; it is not quantified and does not offer predictive validity in time and space. Yet in the 1960s and today, the resonance which the tragedy of the commons has to resource exploitation is striking. Perhaps one of the most effective illustrations of the tragedy is everyday consumer behaviour. A useful example within this framework would be our use of water resources. The summer of 2006 in the UK witnessed a major crisis in water use in the south east of the country. Drought orders, hose pipe bans and other restrictions were placed on households and businesses in many areas of Kent, Sussex and London. In this context, the day of reckoning, according to Hardin's tale, is the day when social stability in the form of economic development means that the state and society can afford to enhance lifestyles with consumer products, all of which consume water. Far beyond the basic necessities of drinking and sanitation, which are held within the limits of resource exploitation, each rational human being asks themselves what the utility of washing their car, watering their garden or filling their swimming pool is, against the potential negative impact. As with the herdsmen, the positive impact is considerable: a clean vehicle, a lush garden and a full pool in which to cool off. In a society 'locked' into consumption that places symbolic value on consumer products and culturally distinguishes between social strata by the consumer 'signs' that are provided by individuals (Bourdieu, 1960), the positive effect of using water to present these signs and symbols of wealth and success are far more preferable than the perceived negative impact of exploitation. Such negative impacts are framed not in relation to overall resource exploitation, but in terms of the economic cost associated with water use. Given the symbolic and sign value associated with what is perceived as water 'need', this economic cost is considered acceptable. In addition, in a nation that is generally considered to be temperate and with high levels of rainfall, resource exploitation is not considered an issue. Indeed, the water crisis of 2006 witnessed a range of discourses concerning water use being employed and developed by a range of consumer groups through the media. Perhaps the most dominant of these was the argument that there was only a water crisis because of the mismanagement of resources by water companies. Their inability to prevent leaks and still make profits was seized upon instead of the actual water use of each individual and how this related to supply.

Such an illustration touches the core of the argument surrounding our green dilemmas; so many of the environmental problems that present themselves are, by definition, the result of *individual* exploitation of resources. Yet they are manifested at the *global level* and thus appear wholly disconnected from everyday lives and practices. Such a disconnection ensures that these green dilemmas are ones which are framed with suspicion and vigorous debate. The key question that is often posed relates to whether any individual's action, in the knowledge that it *does* contribute in some way to exploitation, can have an impact in effecting positive change. The refrain, as we shall see more fully throughout this book, is that 'yes, but only if *everyone* participates'.

This final green dilemma (individual–society) is intimately related to the preceding two dilemmas of conservation–growth and time–space. Each poses critical questions related to how individuals relate to society, how environmental issues are framed and the types of lifestyles individuals wish to lead in these contexts. Having established these three dilemmas, we now turn to an examination of how policy has attempted to deal with these dilemmas. First, a framework for analysing environmental policy is outlined. The discussion then moves to a consideration of the political shifts that have framed the current political discourse of environmental issues. Having mapped out these frameworks for analysis, the chapter closes by introducing the structure of this book.

Environment, Society and Policy

Environmental policy has long acknowledged the implicit link between resolving and ameliorating environmental problems and human behaviour. Yet the 'rendering' of those environmental issues has long been the subject of debate within policy circles. This 'rendering' process (Rose and Miller 1992) refers to the ways in which an entity (such as the environment) is 'rendered thinkable' in policy discourse. For example, in reflecting on rural policy, it could be argued that 'the rural' as an issue of concern to politicians and policy-makers has undergone a series of rendering processes, defined by the dominant policy-related paradigms at the time. Accordingly, rural policy has been framed in a range of different contexts, from agricultural, economic and productive, to a new political context that defines the countryside in terms of a post-productivist and service-based and social context (Wilson 2001). In the same way, the environment has been 'rendered thinkable' through a series of contextual shifts that have enabled environmental issues to be reframed in ecological rather than social or economic contexts.

One clear illustration of this rendering process is the way in which a prominent environmental issue has been framed and reframed in society. Municipal waste was, for most of the twentieth century, a growing environmental problem (Pellow 2003; Weinberg et al. 2002) that reflected society's increasing reliance on consumables. Yet 'waste' as a discourse has been reframed several times.

Discourses of waste in western societies have focused around both negative and positive constituents, with 'waste' regarded as a significant resource in times of hardship, for example during the economic depression of the 1930s or World War II. In contrast, waste has been conceptualised in negative terms as a significant public health issue, and only recently has waste been rendered thinkable as an environmental problem. This example reflects a number of the environmental issues, which in and of themselves have only recently been framed in terms of 'environment'. One of the reasons behind this rendering process is without doubt the political and economic contexts that form the background to environmental thought. However, the post-World War II history of environmentalism can be charted most effectively with reference to some seminal works on environmental issues and the specific problems that they highlighted.

Environment and Society: A Twentieth-century Perspective

To chart the growth of modern environmentalism means to begin with the nation which has recently been pilloried for its lack of environmental commitment. The 1950s and 1960s in the United States represented a significant growth in environmental technologies and the use of agricultural chemicals to enhance yield and control pests. Of the chemicals used, DDT (a white chlorinated hydrocarbon used as an insecticide) was perhaps the most well-known and most effective. Yet the use of chemical fertilisers and biocides for the intensification of agriculture was not without incident. In 1948, the population of Donora, Pennsylvania, had fallen ill when a mysterious sulphurous fog had descended on the settlement. By the late 1960s, concern was rising over the use of artificial chemicals for agriculture. The first major popular text to highlight this issue was Rachel Carson's (1963) *Silent Spring* which provided a fateful account of the impact of DDT. Carson's work begins with an emotive, yet strikingly accurate 'Fable for Tomorrow'. She describes a 'town in the heart of America' where 'all life seemed to live in harmony with its surroundings' (3). This harmony is expressed in purely natural terms, without reference to human presence. She then emotively describes how 'a strange blight crept over the area and everything' (3). In portraying what was to follow, the description shifts from the natural to the human, with the 'blight' encompassing and embodying nature and humans.

Carson's *Fable for Tomorrow* attempted to emphasise the harmful role of DDT both in the landscape and on society. In reaching critical acclaim, it is widely regarded as the first text which highlighted a significant environmental problem within a wider political context. Yet it did much more than merely highlight a specific environmental issue in time and space; Carson's work epitomised the emerging values of a modern society that was beginning to render nature thinkable in intrinsic terms. The closing remarks of her account bring into sharp focus the lines of contention that would and still do form much of the basis for environmental philosophy and value judgments that will be investigated further in Chapter 5 'The "control of nature" is a phrase conceived in arrogance, born of the Neanderthal age of biology and philosophy, when it was supposed that nature exists for the convenience of man'. The fundamentally ecocentric position that Carson adopted was one that others developed, the most notable being Lynn White in her 1967 article in *Science*. *The Historical Roots of Our Ecological Crisis* strongly argued that environmental deterioration in western societies was born out of a fundamentally anthropocentric view of nature-society relations that were grounded in Judeo-Christian belief systems. In brief, White's (1967, 1205) argument posited that 'Christianity, in absolute opposition to ancient paganism and Asia's religions ..., not only established a dualism of man and nature, but also insisted that it is God's will that man exploit nature for his proper ends'. White's assertions can be drawn from books such as Genesis, where clear reference is made to the role of humans within Creation. Perhaps the most familiar passage is with regard to God's intended role of man within nature (Genesis 1, 26–30, New International Version):

> Then God said, 'Let us make man in our image, in our likeness, and let them rule over the fish of the sea and the birds of the air, over the livestock, over all the earth, and over all the creatures that move along the ground.' So God created man in his own image, in the image of God he created him; male and female he created them. God blessed them and said to them, 'Be fruitful and increase in number; fill the earth and subdue it. Rule over the fish of the sea and the birds of the air and over every living creature that moves on the ground.' Then God said, 'I give you every seed-bearing plant on the face of the whole earth and every tree that has fruit with seed in it. They will be yours for food. And to all the beasts of the earth and all the birds of the air and all the creatures that move on the ground – everything that has the breath of life in it – I give every green plant for food.' And it was so.

These Old Testament passages are also reflected by the anthropocentric tones expressed in the New Testament. Again, one of the most frequently quoted passages of New Testament scripture is Jesus' teaching on worry (Luke 12, 22–24, New International Version):

> Then Jesus said to his disciples: 'Therefore I tell you, do not worry about your life, what you will eat; or about your body, what you will wear. Life is more than food, and the body more than clothes. Consider the ravens: They do not sow or reap, they have no storeroom or barn; yet God feeds them. And how much more valuable you are than birds!'

On superficial inspection one might be led to the conclusion that White's thesis was supported by Christian scripture. Yet the re-interpretation of these and other passages is significant. Young's (1990) analysis of these and other New Testament verses can lead to a very different conclusion. He cites the importance of the Parable of the Tenants (or Parable of the Vineyard) to be found recounted in the Gospel's of Matthew, Mark and Luke. In Matthew (21, 33–41, New International Version) the parable reads as follows:

> Listen to another parable: There was a landowner who planted a vineyard. He put a wall around it, dug a winepress in it and built a watchtower. Then he rented the vineyard to some farmers and went away on a journey. When the harvest time approached, he

sent his servants to the tenants to collect his fruit. 'The tenants seized his servants; they beat one, killed another, and stoned a third. Then he sent other servants to them, more than the first time, and the tenants treated them the same way. Last of all, he sent his son to them. 'They will respect my son,' he said. 'But when the tenants saw the son, they said to each other, 'This is the heir. Come, let's kill him and take his inheritance.' So they took him and threw him out of the vineyard and killed him. 'Therefore, when the owner of the vineyard comes, what will he do to those tenants?' 'He will bring those wretches to a wretched end,' they replied, 'and he will rent the vineyard to other tenants, who will give him his share of the crop at harvest time.'

Young (1990) regards these verses as critical to an understanding of God's vision of how humans should be stewards of Creation, not exploiters of it. There is little controversy that this passage still places humanity at the centre of Creation and that God's earth is centred around humans, but Young (1990) rejects White's (1967) argument that the exploitation and environmental degradation of natural resources can be implicitly traced back to Biblical teaching. Indeed, so much of the Old and New Testament's are imbued with references to nature, that to interpret an anthropocentric position as one that justifies and promotes exploitation is questionable. Indeed, as shall be seen later in this text, evidence suggests that those with an active faith are more likely to hold positive views concerning environmental action.

The thesis put forward by White (1967) ignited debate regarding the potential of humans to exploit the earth's resources to such an extent that the term 'environmental crisis' was one that became synonymous the late 1960s. During a period of exponential population growth, the central issue to emerge was the extent to which a population growth rate could be sustained by the finite resources provided by the planet. Central to this discourse of finite resources was what became known as the 'limits to growth' debate. The evidence on which such a debate was focused can be categorised both in terms of a popular discourse that emphasised the seemingly ever-increasing levels of resource exploitation evident in western societies, and a more theoretically grounded notion of resource exploitation.

The Limits to Growth

Attempts to model the relationships between resource exploitation, economic output and development commenced during the nineteenth century but it was one dour cleric who became dominant within the academic discourse at the time. The thesis put forward by the eighteenth-century clergyman and demographer Thomas Malthus was to spawn what became known popularly as 'the dismal science', for reasons which will become apparent. *An Essay on the Principle of Population* (1798) had as its central argument the notion that geometric (or what we now term exponential) population growth could not be sustained by food production. Malthus envisaged a cycle of population growth that began with what he termed the introduction of 'alien technology'. This would have referred to emerging farming techniques that could increase yield, but alongside this illustration, one could easily refer to any number of technological innovations that have enhanced food production and distribution. Such advances stimulate population growth, as production in the first instance can meet demand and provide a surplus. However, Malthus calculated that population growth was exponential in nature, whereas food production was only likely to increase in a linear fashion. The assumption underlying this hypothesis was the additional welfare and health benefits afforded initially by a wealthier and prosperous society, enabling not only birth rates to increase but mortality rates to decrease. Using a series of 25-year cycles to illustrate his point, Malthus concluded that over 75 years, food production would increase enough only to be able to feed under half of the likely population. Malthus' conclusion was that the discrepancy between food production and population growth would be addressed by famine, leading to a decline in population to sustainable levels. The general principle of Malthus is illustrated in Figure 1.2. The 'cycle' begins with the application of 'alien technology', which refers to new intensive and mechanised agricultural methods. Alongside improved productivity, the growth of the economy permits enhanced health and social welfare. The two factors combine to produce population growth. Using the formula outlined above, population soon begins to outstrip food production and standards of living begin to fall, although fertility rates remain high. The widening gap between food production and population leads to 'misery, disease and famine' (a clear indication of why this was termed 'the dismal science') and accordingly population falls. The cycle then begins again. This apocalyptic vision of the consequences of population growth and subsequent collapse were to be revisited in what became the most seminal piece of scientific work on the limits to growth debate.

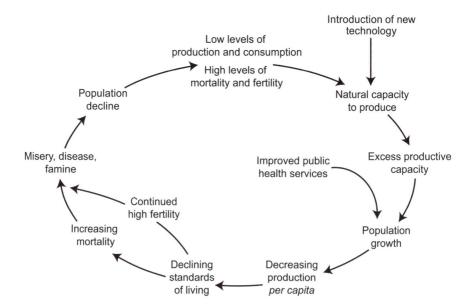

Figure 1.2   The Malthusian cycle of population growth and decline

In 1968 30 individuals from the scientific, education, humanist and industrial fields met in Rome as an informal gathering to discuss what the preface to their final report describes as 'the predicament of mankind'. As a 'virtual college', the Club of Rome grew significantly in the late 1960s as interest grew in this ambitious project. The Club is most well-known for their work on limits, but the project on which they embarked was far more ambitious, including an investigation of the loss of faith in institutions, urban sprawl, unemployment and the alienation of youth and degradation of traditional values. Yet the report came to be defined by its focus on the degradation of the natural environment and the ability of humans to cope with such damage. In what the final report termed 'modelling the human condition' (Meadows et al. 1972), the aim of the research was to provide a predictive tool based on the essential assumptions of Malthus regarding food production and population growth, but with other key variables included. These related to industrialisation and therefore the rate at which resources were consumed; the amount of renewable/non-renewable resources consumed; and finally pollution. The final variable was key to the arguments of the Club of Rome, implying that there was an implicit link between industrial pollution and food production.

Two major scenarios were modelled by the Club, although numerous models were developed. The standard scenario plotted population growth against the other variables from 1900 to 2100, using historical data from 1900 to 1970. Figure 1.3 shows this reference case scenario. The clear assumption provided in the model relates to resource depletion, which is pegged to industrialisation. These are clearly related to food production, which shows a significant decrease in line with these variables. Population continues to grow until the death rate is radically increased by a combination of food production decreases and medical service reductions.

This reference case scenario was added to by a major adjustment in the assumptions of this model. To take into account the possibility that the Club of Rome had under-estimated the potential natural resource base, the amount of such resources were doubled (Figure 1.4). It is using this case that the importance of pollution was emphasised by the modellers. Meadows et al. argued that with increased natural resources came enhanced exploitation and therefore higher levels of pollution, released at an alarming rate. The impact of this pollution would be twofold. In the first instance, the death rate was directly increased in line with pollution. However, food production also collapsed due to the excessive pollutants emitted from major industrialisation. In this scenario, the collapse of the population was more immediate than in the reference case, due mainly to the significant impact of environmental pollution.

Both the scenarios provided the same conclusion: a decreasing level of food production, depletion of resources and subsequent population decline. The report of the Club of Rome ends with this sobering call:

> The last thought we wish to offer is that man must explore himself – his goals and his values – as much as the world he seeks to change. The dedication to both tasks must be unending. The crux of the matter is not only whether the human species will

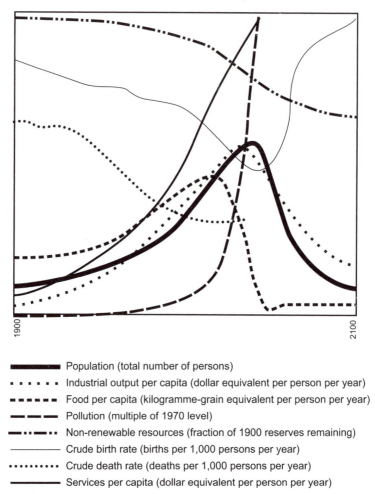

Population (total number of persons)

Industrial output per capita (dollar equivalent per person per year)

Food per capita (kilogramme-grain equivalent per person per year)

Pollution (multiple of 1970 level)

Non-renewable resources (fraction of 1900 reserves remaining)

Crude birth rate (births per 1,000 persons per year)

Crude death rate (deaths per 1,000 persons per year)

Services per capita (dollar equivalent per person per year)

Figure 1.3 'World model standard run' from *The Limits to Growth* (after Meadows et al. (1972). Reproduced with the kind permission of Professor Dennis Meadows

Source: Meadows et al. (1972).

survive, but even more whether it can survive without falling into a state of worthless existence. (Meadows et al. 1972, 197)

This final call for dialogue opens a further chapter that will be examined later in this text, but the crucial point made by the Club of Rome was that humankind faced a predicament between continuing economic prosperity or environmental catastrophe. As demonstrated both by popular and academic discourse, this predicament can be framed according to the temporal and spatial contexts in which individuals live and the consequences of the consumer society.

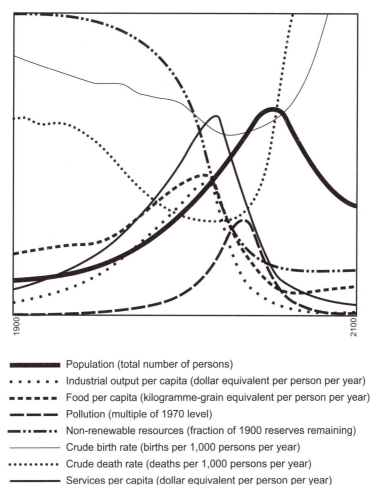

Population (total number of persons)

•••••• Industrial output per capita (dollar equivalent per person per year)

▪▪▪▪▪▪ Food per capita (kilogramme-grain equivalent per person per year)

— — — Pollution (multiple of 1970 level)

▬••▬•• Non-renewable resources (fraction of 1900 reserves remaining)

——————— Crude birth rate (births per 1,000 persons per year)

•••••••••• Crude death rate (deaths per 1,000 persons per year)

——————— Services per capita (dollar equivalent per person per year)

Figure 1.4   Scenario from *The Limits to Growth* assuming a doubling of natural resource reserves (after Meadows et al. 1972). Reproduced with the kind permission of Professor Dennis Meadows

Source: Meadows et al. (1972).

Criticisms of The Limits

Perhaps the most sustained critique of *The Limits* was of a philosophical nature. It was argued that *The Limits* proposed a fundamentally anti-growth agenda, in which the solution to environmental destruction was to reduce growth in both the developed and developing world (McCormick 1989). As McCormick (1989) demonstrates, such a position was utilised widely by both academics and politicians to weaken the arguments of Meadows and his team. From one perspective, the 'anti-growth' agenda was interpreted in a wider social context as a vehicle for politicians opposed to liberal social developments to put forward their traditionalist views. From a Marxist perspective, *The Limits* was conceived as a document that sought to break working-class resistance to government and authority. Yet overwhelmingly, the document was considered a 'doomsday' assessment; an exercise in negativity that did not permit the possibility that humans could generate solutions to these daunting problems.

These philosophical and conceptual difficulties of *The Limits* were compounded by a greater systematic analysis provided by British academics. Researchers from the Science Policy Research Unit at the University of Sussex produced a series of 13 essays aimed at questioning the technical bases of the report. As McCormick (1989) notes, the British academics were careful not to criticise the overall impact of *The Limits*. They acknowledged that the media coverage and serious consideration by politicians should not be ignored, but highlighted the science behind the predictions. Overall, the predictions were based on a lack of empirical data and, as with all models, had a bounded nature which meant that the transference from theory to practice was likely to be poor. The confidence with which Meadows et al. had presented their models asserted that the predictions they contained were both accurate and probable. In reality, they did provide a stark warning to politicians, but they could not be used, ultimately, as a predictive tool.

Environment and Society: Sustainability, Policy and the Citizen

The predicament posed by organisations such as the Club of Rome appeared to make the future of human progress an issue of 'growth' or 'limits'. These oppositional discourses presented what was undoubtedly a somewhat simplistic notion of the future of humankind. Either growth continued with a 'business as usual' approach, which led to population collapse, or there would need to be a major reduction in resource use, industrialisation and pollution. Such debates were presented as win/lose situations, with no middle ground. Similarly, as early texts on the issue make clear (e.g. O'Riordan 1976), the link between environment and society was considered in oppositional terms. O'Riordan (1976) highlights the role of protest as a means of expressing environmental consciousness in the 1960s and 1970s. Indeed, this was the era in which many of the well-known environmental pressure groups, such as Greenpeace and Fiends of the Earth, were established.

This book is positioned at the interface between the emerging discourses of sustainable development, participatory democracy and environmental activism, all of which mesh to provide new perspective on the role of citizens in the environment – society debate. Having positioned the book within the 'green dilemmas', which emphasise the complexities of human agency within the context of time, space and institutions, the book moves on to discuss the notion of sustainable development through the lens of two critical contexts: sustainability and policy (Part 1: 'Contexts'). The period immediately after the publication of the Club of Rome's report was characterised by a series of politically high-profile conferences and meetings that sought to tackle the twinned problems of environmental 'destruction' and population 'explosion'. Led partly by the *Man and the Biosphere* programme of the United Nations, the movement to appreciate the mechanisms by which humans interacted with the natural environment was eventually transformed into what has been the most influential commission convened in recent times. Gro Harlem Brundtland's World Commission on the Environment and Development took four years to report after extensive work. Their report (UNCED 1987) provided the catalyst for the 1992 World Conference on Environment and Development (IUNCED 1992), more commonly known as the Rio Earth Summit. This conference set the agenda for the institutionalisation of sustainable development. Yet as will be seen in Chapter 2, what Brundtland envisaged as sustainable development has been the subject of considerable theoretical debate. Sustainability as a concept is deceivingly simple, yet practically complex. In one sense, it is eminently achievable and yet in other ways it appears elusive. It can appear to be all-encompassing and yet in being so, can encompass no-one. Chapter 2 therefore explores how the attempts of academics and policy-makers to provide an acceptable form of environmental protection within the context of a developmental discourse has fared in a world that in many ways is defined by growth. In particular, interpretations of sustainable development will be explored, which involve ways of defining different forms of capital and how we can conceive of 'weak' and 'strong' approaches to sustainable development. All of these arguments contribute to an understanding of how the shift from 'environmental crisis' to the political mainstreaming of sustainability has been a process which has been reflected in the growing importance of citizens and participatory discourses in the environmental arena. The chapter will also provide a critical device for framing sustainability, according to the various principles, concepts and applications of sustainability. This will provide a framework for appreciating the forthcoming chapters and the alternative approaches which have been developed to explore sustainability and the citizen.

Chapter 3 places the conceptualisation of sustainability within a political and institutional context in the UK and illustrates the role that different institutions play in promoting citizen engagement in sustainability. It examines the response to the Brundtland Commission's proposals by the then-Conservative government and their response to the Rio Earth Summit. Through an analysis of national policy documents, it will plot the phases of sustainable development policy in the UK to the present, illustrating the increasing importance of citizen engagement for sustainability and the shifts in policy to reflect the significance of behaviour change to sustainability. Alongside this chronological analysis, examination will seek to appreciate sustainable development policy both in terms of scales and sectors. Geographically, sustainable development in the UK has permeated every level of both government and non-governmental organisations and this has clear implications for the range of participatory mechanisms developed for citizen engagement. What emerges from this analysis is a diverse picture, characterised by significantly varying levels of progress and commitment and, without doubt, a large divergence between rhetoric (policy) and reality (practice).

Part 2 ('Perspectives') of the book focuses on individual action for sustainable development and seeks to conceptualise how behaviour for sustainability is framed within western societies. In Chapter 4, three approaches to examining the nature of behaviour for sustainable development are outlined. In the first instance, the approach that has been adopted by many policy-makers is outlined, which seeks to use awareness raising techniques to change behaviour. This perspective on behaviour and behaviour change is driven by a belief that positive steps towards a more sustainable citizenry are blocked by a lack of awareness and information. Filling this knowledge 'gap' is therefore necessary to effect change. However, the central assumption is that individuals are predisposed to help the environment and that it is a lack of information that prevents them from doing so. This approach can be termed the 'information deficit' model and illustrated by using four well-known examples from national behaviour-change campaigns.

Chapter 4 then provides a critical analysis of this perspective through the lens of two alternative approaches. The first, which is commonly known as the deliberative or 'civic' approach, is derived from a critical analysis by geographers of the 'rationalistic' (Owens 2000) approach applied by policy-makers. The 'civic' approach attempts to understand behaviour change in terms of the institutional and social contexts in which individuals are placed. Central to this thesis of behaviour change is that the isolation of individuals from the environmental problems which surround them, and an increasing emphasis on scientific, rather than lay sources of information, ensures that a disconnection occurs between the natural and the human. To this end, behaviour change is regarded as reliant on engagement of citizens in both defining and resolving environmental issues.

The second alternative to the information deficit model and the approach adopted in this book is to develop a greater appreciation of the factors that influence behaviour change and to examine this within the context of lifestyles. This perspective argues that current policies for behaviour change are focused on encouraging activities that do not relate to the lived experiences of individuals. Indeed, it is argued that policy has failed in part because of the sweeping approach it has taken to encouraging behaviour, which has used media campaigns focused on the population in general. Such campaigns by definition do not allow messages to be tailored to specific lifestyle groups who may have particular behavioural characteristics. Accordingly, Chapter 4 concludes by arguing that behaviour change is only likely to occur when we have a fuller understanding of individuals' actions and can apply this to specific segments within society. To assist in an understanding of behaviour and to frame the debate for the remainder of the book, a conceptual framework is outlined that seeks to appreciate the relationships that exist between alternative influences on behaviour.

Chapter 5 then critically explores the notion of environmental action – or what it means to be 'green'. A series of well-known 'actions' to help the environment are outlined within the context of existing policy for behaviour change. Using these as a working definition of environmental action, the chapter moves on to consider the factors that researchers have argued influence different forms of environmental behaviour, using the framework from Chapter 4 as a guide and begins by examining the role of values in constructing environmental lifestyles.

As underlying principles in individuals' lives, values constitute an important factor that can impact on decisions relating to the environment. This is also the case with regard to two other key components: situational characteristics and psychological factors. Considerable research suggests that an individual's social context, not least their demographic profile, can have a significant impact on their environmental action. Indeed, the attitudes they hold towards such behaviours will also have a bearing on behaviour. Using these three groupings to organise the debate, the chapter ends with a description of how these factors can be examined within a case study approach to exploring environmental action.

Accordingly, in the first chapter of Part 3 ('Approaches') Chapter 6 examines the nature of environmental action using the vehicle of research undertaken in Devon in 2002, part of a large Economic and Social Research (ESRC) project. In the first of four analytical chapters exploring the nature of environmental action in the UK in the twenty-first century, Chapter 6 explores the activities that individuals reported participating in and what these levels of activity can tell us about contemporary lifestyles. In particular, the chapter explores how environmental action can be related to everyday practices of individuals. It will examine the alternative behavioural practices that can be identified and related to everyday practices and will describe three dominant behavioural themes, related to consumption, habit and waste management. The chapter concludes by challenging the conventional notion advanced by policy-makers that environmental action should be promoted according to problem-based activities, rather than behaviours focussed around existing daily practices.

Chapters 7 and 8 use these behavioural classifications to examine how the variables outlined in the theoretical framework of behaviour interact to provide a basis for appreciating environmental action. In Chapter 7, a series of cluster analyses are used to explore the potential for creating lifestyle groups based on the environmental behaviour reported by individuals. These groups are then explored to examine the diversity of the attitudes and socio-demographics illustrated by the lifestyle segments. The use of segmentation as a practical technique to explore the lifestyle attributes of individuals displaying alternative behavioural properties is explored and justified in terms of the potential that can be offered by exploring the definitive characteristics of these groups in terms of their values, situation and psychological variables. The chapter concludes by arguing that behaviour change policies can be greatly assisted by the use of classification of individuals.

In Chapter 8, a series of multiple regression analyses are used to present three models of environmental action relating to the behavioural classifications described above. As will be seen, depending on the type of behaviour in question, the variations between the role of values, situation and psychological factors are significant. The chapter concludes by arguing that a more effective approach to behaviour change is likely to be derived from appreciating the alternative influences of values situation and psychology on different types of behaviour.

Finally in Part 4 ('Applications'), these approaches are placed into a practical context with the exploration of the role social marketing can play in behaviour change. This is undertaken in Chapter 9 using data gathered from subsequent focus group work as part of a DEFRA-funded research project which sought to extend the previous work for ESRC and examine the potential for using positive socially-based marketing for behaviour change amongst distinct lifestyle segments. This evidently relates to how policy-makers and practitioners view behaviour. It will be argued firstly that behaviour needs to be contextualised within everyday practices and therefore embedded within the social and cultural networks that individuals use to structure their daily lives. Secondly, the social and cultural domains which different everyday practices inhabit are underlain and driven by alternative factors, which necessitate policy-makers to undertake a different approach to changing such behaviour. Finally, it will be argued that policy needs to focus on specific lifestyle groups that demonstrate specific behavioural characteristics. These lifestyle groups will hold alternative values, attitudes and are likely to be representative of different social groups.

These three guiding principles, it will be argued, are the basis for a step-change in the ways in which we view individual's behaviour and encourage real shifts in attitudes and behaviour. However, the text does not provide a 'fix all' solution or a tool kit for success, rather it points to key areas that require attention and debate. At the centre, however, is the clear notion that helping the environment is not merely a simple case of changing behaviour; it is a decision to change a lifestyle, a way of thinking and working, that necessitates a significant commitment.

Accordingly, in Chapter 10, the three elements of the text (contexts, perspectives and approaches) are drawn together by revisiting the three green dilemmas explored in this initial chapter. As the reader will become aware, this book seeks to explore these dilemmas through the lens of individual commitments to the environment, exploring the conflicts experienced by citizens as they negotiate the concept of sustainability. Such negotiations involve a complex meshing of the three dilemmas: the decision to have a lifestyle characterised and driven by economic imperatives, or a conscious rejection of this approach towards reduced personal consumption and greater conservation; the tension between decisions taken at an individual scale at one point in time and their impact globally in the future; and the conundrum of seemingly limitless *personal* consumption alongside evidence that suggests that sustainability may rely on a more equal sharing of 'the Commons'. Such dilemmas are evidently socially constructed and a glance at the popular media reinforces the potential for classifying such conflicts in strict binary terms. Indeed, depending on one's perspective, such extreme positions can be heavily critiqued scientifically. Yet the focus of this book is to consider how such dilemmas, whether real or constructed, are framed and reflected in personal commitments to the environment and how policy-makers can deal with these messages more effectively, providing a route to sustainability that presents it as a desired process.

Chapter 2

Sustainability

The New Politics of Sustainability

There can be little doubt that the concept of what we have come to understand as 'sustainability' is the dominant political discourse of the early twenty-first century, penetrating numerous sectors of society, from academic disciplines, political parties and government agencies, to local authorities and community organisations. Accordingly, as an over-arching discourse, sustainability has permeated academic and socio-political networks both vertically and horizontally. Academically, sustainability has and continues to present significant challenges to research and teaching. Disciplinary boundaries do not easily lend themselves to the study of sustainability, with both the natural and social sciences sharing intellectual capital, yet seemingly incapable of providing a realistic interdisciplinary agenda. Politically, sustainable development poses significant challenges for integrating such a 'slippery' concept into practice and arguably presents the most pressing difficulties for politicians trying to reconcile environmental, economic and social agendas. However, a key question is why sustainability has been so difficult to mould into a single and universally accepted discourse. The lack of any single definition and the resulting multitude of conceptual models, frameworks and policies ensures that sustainability has and continues to emerge as a concept to which most can pledge allegiance at one level, yet still hold radically different opinions on a range of social, economic and environmental issues at another level. From the perspective of citizen engagement, these conceptual differences do not create a simple binary between 'good' and 'bad' practices for sustainability, but rather a complex and overlapping set of discourses which often reflect contradictory positions.

To appreciate both the historical roots of this contemporary conceptual position and to create a framework for analysing sustainability within the context of citizen engagement, this chapter first explores the historical antecedents of 'sustainable development' before 1987 and in particular the emergence of a politics which attempted to synthesise 'environment' and 'development'. Such a politics has radically altered the relationship between citizens and environmental policy, with the recognition that sustainability needed to become an inclusive concept, reflecting the needs and desires of society and individual people alongside environmental protection and macro-economic growth. Having established the key events in embedding sustainability as a mainstream political discourse, the chapter then explores the contested nature of sustainability, examining a series of frameworks for defining and practicing sustainable development. Again, such frameworks illustrate the complex notion of sustainability within the context of citizen engagement. However, it will be argued that sustainability must be analysed on a number of different levels, from principles to concepts and then to application. It will illustrated how such a framework for analysis reveals the multiple routes to sustainability and the importance of reflecting on the contingent nature of knowledge and understanding when exploring sustainability. In doing so, the chapter seeks to provide a link between the political and definitional debates surrounding sustainability and therefore to illustrate the alternative lenses through which to view notions of citizen engagement for sustainability.

The Political Roots of Sustainable Development

The publication of *The Limits to Growth* (Meadows et al. 1972) was the zenith of what has been described as the 'Environmental Crisis' (Chapter 1). In an early age of computer simulation, reliant on the assumptions provided by Malthusian population dynamics, the predictions were considered with grave concern. The quantitative revolution of the 1960s also ensured that the data presented were considered within the broader context of a widely held belief that quantitative approaches to prediction were both valid and reliable. Indeed, the Club of Rome that had sponsored the work provided explosive potential because of their academic respectability and neutrality. Thus the scene was set for a decade in which the major axis for debate focused around the ways in which society could tackle a major set of environmental crises, whilst acknowledging the role which exponential economic development and population growth were playing in exacerbating such problems. As will be illustrated in the following paragraphs, this was the decade characterised by high-level political discourse, reflecting the notion that environmental issues were a global problem, meriting global and national solutions, a position which on the most part excluded any discourse framed around the role of individual citizens.

Accordingly, perhaps the most important political impact of the *Limits to Growth* was to focus the minds of delegates to an international conference being held in Stockholm in 1972. Stockholm is regarded as being the most significant international meeting in terms of institutionalising environmental issues at a political level (McCormick 1989). Yet it was the earlier Paris Biosphere Conference in 1968 which provided much of the basis for the workings of the Stockholm meeting. This scientific meeting held in the French capital focused around the interactions between humans and the lithosphere, hydrosphere and atmosphere. Quite apart from the 20 recommendations that emanated from the meeting, the overall conclusion of the delegates was that an increased level of understanding and therefore research was required to appreciate the complex interrelationships between humans and nature. One of the most well-known of these research programmes was Man and the Biosphere (MAB). This intergovernmental and interdisciplinary research programme had four main aims, focused around the identifications and assessment of biospheric changes related to human behaviour, the study of interrelationships between natural processes and social and economic processes, the measurement of change and complexity in natural processes and, related to this theme, the promotion of coherence in global research on the environment.

To a far greater fanfare and involving the political as well as scientific elites, the Stockholm conference was an attempt by the United Nations to approach the challenges of environmental problems from the perspective of a scientific understanding of ecological processes. Once again, the influence of a purely quantitative and logically positivist view to investigating the relationships between nature and society was dominant (Moffatt 1996). This is illustrated by the precursors to the Stockholm meeting, which were two United Nation's led programmes on the International Biological Programme (IBP) and the International Geosphere Biosphere Programme (IGBP). Both of these programmes, the latter succeeding the IBP, were developed during the 1960s as a means by which to examine the biological bases of human economic development and welfare. Each programme used a prescriptive scientific operationalisation. This was focused around ecological problem identification, followed by the convening of a team of scientists to address the issue and write an action plan. This action plan was then operationalised and evaluated.

The role of the IBP and IGBP was significant, as it reinforced the principles embedded in theoretical work such as Hardin's (1968) *Tragedy of the Commons* and the implicit assumptions underlying Malthus' (1798) beliefs regarding the population's reliance on ecological stability for growth. Indeed, the reader should also note that the investigation of relationships and reliabilities between ecology and economy lay at the heart of the IBP and IGBP. Questions of social welfare and the role that individuals played in framing and influencing environmental issues were not included in the programmes.

Accordingly, the Stockholm Conference was convened in the light of the essential assumption that the link between environment and economy could not be overlooked. The meeting comprised representatives from 113 nations, 19 intergovernmental agencies and 400 non-governmental organisations. Perhaps the most significant political dimension of this meeting was the emphasis placed by developing nations on their specific social and economic needs. Strong (1977, 165) has noted that '... the concern of the environment which led to convening of the Stockholm Conference came almost exclusively from the industrialised countries'. As Strong goes on to note, these concerns were related to the pollution of resources such as air and water and the depletion of wildlife. Yet the developing nations were only inclined to take Stockholm seriously if issues of development could be examined within what many in the developing world considered an aesthetic concern. To do this, it would be necessary to include issues such as loss of productive soil, increased levels of desertification and the management of tropical forest ecosystems. Through the vehicle of the United Nations General Assembly, the developing nations were able to exert the necessary pressure and Stockholm became a much broader meeting than had initially been conceived.

Stockholm produced a declaration, a set of principles and an action plan. The declaration was not legally binding, but rather an attempt to provide a statement of clear intent that all nations could take back to their own administrations. Stockholm's major impacts were to raise the profile of the environment on a world stage and to foster greater appreciation by the developed world of the needs of the developing world. The inclusion of so many NGOs also ensured that Stockholm became a truly inclusive meeting.

The legacy of the Stockholm Conference was significant. Based on the Action Plan produced at the 1972 meeting, the United Nations formed the basis of the United Nations Environment Programme (UNEP). Still active today, this programme was designed to provide the basis for a UN voice for the environment. Through environmental assessment, management and supportive measures, the programme would provide the conduit for UN-sponsored environmental programmes.

The Road to Brundtland

Although the United Nations Environment Programme provided a conduit for environmental initiatives spearheaded by the United Nations, logistical and organisational problems often meant that its efficacy was somewhat diluted. Indeed, throughout the 1970s a shift occurred that led to greater emphasis being placed on the relationships between environment and development, at the expense of a more orthodox and technocratic interpretation of environmental issues. There were assertions that the technical and 'top-down' solutions being proposed by the UNEP and other organisations were unlikely to have a significant impact in either their understanding or resolutions of environmental problems. A greater emphasis needed to be placed on societies and individuals. The interpretation of key environmental problems, such as desertification, could not, it was argued, be understood merely in terms of a natural process and specific and prescriptive ameliorative measures. Rather, an appreciation of the interlinked processes of environmental change and socio-economic development needed to be appreciated. The solutions needed to be the result of a more democratic process. Indeed, such solutions needed to involve indigenous populations whose local knowledge could play a vital role in resolving environmental issues that technical and scientific approaches were only likely to exacerbate.

Central to the Paris Conference on the Biosphere in 1968 and to a lesser extent, the Stockholm Conference in 1972, were issues of how to accurately define the environment and more crucially, environmental problems. In the early 1970s, the approach adopted by policy-makers was focused significantly on the natural environment and a clear separation was defined between humans and nature. The links and relationships between natural process and human development were considered to be one-way and uni-definitional in nature. This implied, particularly from a developed-nation perspective, that environmental issues were a simple result of cause and effect. Industrial processes, for example, were the cause of problems such as acid rain and smog, which could be addressed by greater efficiency. Yet the increased presence and persistence of developing nations at the Stockholm Conference and within the UNEP ensured that environmental problems became considered as part of a wider system, where a multi-layered and multi-directional set of processes and relationships could be envisaged between environmental and socio-economic processes.

A second shift related to what can be termed 'methodological' issues. The Man and Biosphere Programme (MAB) emphasised the scientific basis for understanding environmental issues. Couched within the quantitative revolution, the assumption that natural science could provide both an analysis of the symptoms and a prescription to resolve environmental problems was significant because it set the research agenda on environmental issues for over a decade. The attempts by the Club of Rome to model the 'human condition' are indicative of such methodological approaches. Yet they could only use the data available and inevitably overlooked a host of factors that would become significant in understanding the basis for environmental degradation. By the early 1980s, however, it was acknowledged that a greater reliance on social scientific approaches was required. The growth in social science disciplines concerned with environmental issues is illustrated by the development of subdisciplines such as environmental sociology, environmental psychology, environmental geography and environmental economics. Each of these subdisciplines sought to provide a perspective on environmental issues from their own understanding. Once again, such developments represent a shift from a technocratic model of researching environmental problems to a wider consideration of social and economic factors.

Finally, the climate of 'political solutions' also shifted from the 1970s to the 1980s. The early 1970s were characterised by a high-level international debate relating to two distinct and mutually exclusive constituencies relating to how environmental problems could be tackled. These two positions were embodied by two men: Paul Ehrlich and Barry Commoner (McCormick 1989). Ehrlich was a neo-Malthusian who was characterised in the popular media as an alarmist. He warned that unless population growth was controlled, starvation and plagues were the destiny of humankind. The only solution to this 'inevitable' outcome was the restriction of population growth and economic development. This polarised position was countered by Professor Barry Commoner of Washington University. He argued that whilst economic growth and population increases inevitably had impacts on the environment, the major threat to the environment lay in the nature of this growth. He asserted that economic growth had been undertaken inefficiently and that it was not our use of resources that was problematic, but rather the way in which these were utilised. In his terms, this was in an inefficient and wasteful manner. Greater efficiency and investment in technology were therefore the solutions to the environmental crisis.

This dichotomy between policy approaches is in sharp contrast to what emerged in the 1980s. In the first instance, the high-level nature of the issues encapsulated by the Ehrlich-Commoner debates was replaced by an acknowledgement that political solutions lay in the contextualised and localised settings in which environmental problems were manifested. Second, the stark differences between growth and no-growth were replaced by an acceptance that resolutions to environmental problems were more likely to be accepted if they were perceived as being incremental, rather than radical, in nature. Finally, the political reality of ameliorating environmental problems was brought into greater focus in the 1980s after the apocalyptic predictions of the *Limits to Growth* had died down. The political reality was clearly that environmental policy had to be developed within the context of a growth-orientated society, which would not accept the possibility of reducing standards of living.

These three axes – definitions, methodology and politics – altered to the extent that in the early 1980s the United Nations embarked on what would be the most influential and long-lasting environmental programme it had ever undertaken. In 1983, Gro Harlem Brundtland, former Prime Minster of Norway, was charged with the task of undertaking a four year research project to examine what Meadows et al. (1972) would have termed the 'human condition'. However, this study was to define the 'condition' in very different terms, using alternative methods to answer the question and create the political framework to build a set of policies that would be much longer-lasting than the *Limits to Growth*. It is to this study we now turn our attention.

The Brundtland Commission

Brundtland (WCED 1987) acknowledged that neither she nor her team of researchers could have estimated the breadth of the work that they were to undertake between 1983 and 1987. The four main foci of the Commission related to (WCED 1987):

1) to propose long-term strategies for environmental management in the twenty-first century;
2) to provide mechanisms for greater cooperation and understanding between nations at different stages of economic development and to appreciate the links between individuals, resources, environment and development;
3) to examine how international communalities could more effectively deal with environmental problems;
4) to provide a set of shared aspirations for the expectations and goals for the global community.

The Commission undertook its work in accordance with the mandate provided by the United Nations General Assembly Resolution 38/161. The individuals comprising the Commission were truly international in nature, with representatives from a spectrum of nations who at the time could be classified as first, second and third world. The Commission were involved in taking evidence from five continents, alongside site visits and public hearings, at which over 500 written submissions were received and 10,000 pages of material were collated (WCED 1987). The Commission also appointed three expert panels on industry, energy and food security.

The outcome of the WCED's efforts was a 383-page report that was structured around a diagnosis, analysis and prescription of the issues facing the planet at the time. What is most significant about the report is the title of the first chapter:

'From One Earth to One World'. This phrase succinctly encapsulates the nature of the report, which sought to appreciate the 'interlocking' nature of environmental developmental problems. It also points to what Adams (2003, 88) has argued is a fundamentally 'technocentrist' approach to sustainable development, emphasising the human-centred or anthropocentric nature of Brundtland's assumptions and interests (Moffatt 1996). This is epitomised by the most famous sentence from the report (WCED 1987, 43) which seeks to define 'sustainable development' as '… meeting the needs of the present without compromising the needs of future generations to meet their own needs'.

By placing the notion of human need at the centre of what would become sustainable development, Brundtland set a course which most politicians are still following. Developing this notion of human need, Brundtland (WCED 1987) dealt with three major issues: 'Common Concerns', 'Common Challenges' and 'Common Endeavours'. Without doubt the major concern she and her team highlighted was the paradox (Williams and Millington 2004) between the successes of development and the consequences of such progress on both the human and natural environment. As Brundtland states (WCED 1987, 2), 'Those who look for signs of success can find many', yet these successes and the processes leading to them 'have given rise to trends that the planet and its people cannot long bear'. This paradox is ascribed to the nature of 'interlocking crises' (4) that began to develop. Until the twentieth century, environmental and developmental issues were mainly the preserve and concern of nation states and their impacts remained largely within territorial borders. During the twentieth century, Brundtland (WCED 1987, 4) argues that 'These compartments have begun to dissolve'. The increasingly globalised nature of the world economy has meant that the impacts of development can be seen worldwide and can shift with increasing frequency. Indeed, the global economic system that has encouraged poorer nations to take out large loans for major infrastructure and pump-priming projects has also been the system that has allowed debt to such nations to grow exponentially, forcing over-exploitation of natural resources, reducing the resource base for future generations.

A further concern raised by Brundtland was the lack of international cooperation and organisational ability to deal effectively with the challenges that lay ahead. Brundtland argued that the United Nations needed a stronger role in coordinating a response to the crises that were emerging. Indeed, a further issue of concern related to the unwillingness of institutions charged with environmental protection actively to prosecute those who damaged the environment.

Building on these three concerns (the environment – development 'paradox'; the interlocking nature of crises; and the institutional weaknesses in the international economic and political systems), Brundtland's team turned their attention to a series of 'common challenges'. These began with what the reader will recall initiated much of the environmental debate in the late 1960s – population growth. The arguments of authors such as Meadows and Ehrlich had centred around a neo-Malthusian interpretation of population dynamics, where increases in population led to food scarcity and population collapse. Brundtland tacitly acknowledged that 'Present rates of population growth cannot continue'

(WCED 1987, 95). Yet her team's argument pursued a different line from the neo-Malthusian alarmists of the 1970s by stating that '… threats to sustainable use of resources come as much from inequalities of people's access to resources and from the ways in which they use them as from the sheer numbers of people'. This perspective on population is also matched by the Commission's arguments relating to food security, ecosystems preservation, energy, industry and urban challenges. In each case, Brundtland argued that humans have the ability to overcome many of the challenges that were faced and that a combination of innovation, cooperation and technical developments could be used to ensure that both the natural and human environment were protected.

The final section of the report by WCED set out proposals for what Brundtland termed 'Managing the Global Commons', which sought to demonstrate the viability of international cooperation on environmental and ecological issues. She also emphasised the role of political processes to the success of achieving this goal, highlighting the negative impact of conflict and the 'arms culture' on development and environmental problems. In summary, the Commission provided the framework for further action, including legal measures, which was eventually to form the basis for the action plan for sustainable development, *Agenda 21*.

The UNCED and Agenda 21

In 1989, the United Nations launched the United Nations Conference on Environment and Development (UNCED). As a direct response to *Our Common Future*, this Conference would attract considerable public attention, but most of the outcomes in the final communiqué from this event (UNCED 1992) were undertaken at preparatory meetings before the Conference. As Connelly and Smith (2003) note, the main area for debate before the main Conference in 1992 was *emphasis*. What became known as countries from the 'North', wished to highlight the role of environmental issues that they perceived to be of a short-term nature, directly as the result of technological inadequacies. The 'South', in contrast, argued that such apparently short-term and easily explicable problems were underlain by endemic challenges such as poverty, uneven development and unfair treatment from the North. Accordingly, they made powerful arguments that the remit of the Conference should be centred around such challenges, with a focus on trade, debt, trans-national corporations and poverty. It is without doubt that the developing nations of the South did make headway in this area, with many chapters of Agenda 21 focussing on broader social and economic issues.

At the meeting itself in the Brazilian city of Rio de Janeiro, 176 nations were represented, with a significant presence from non-governmental organisations and environmental groups. The major outcome was the *Agenda 21* document. This communiqué, comprising over 30 chapters on numerous aspects of human development, economic activity and environmental sustainability, was published four months after the meeting in Rio and was itself representative of the huge diplomatic efforts begun in 1989.

Rio did record more outputs than *Agenda 21*, although this is the most lasting and most often quoted document. The Rio Earth Summit, despite the political horse-trading that had gone on in the preceding two years, did produce what became known as the 'Rio Declaration' (Baker 2004), which was referred to at the time as the Twenty Seven Principles of Environment and Development. These covered the four major areas of environmental, economic, social and peace principles. Moffatt (1996) provides a critique of these principles, pointing to inconsistencies not least with regard to how some 'principles' are statements of legal intent rather than based on sound ethical grounds. Nonetheless, what is most intriguing about the principles is the reliance on environmental aspects – twelve of the twenty seven are environmental, demonstrating the clear focus that Rio had on this area. The UNCED also produced the United Nations Framework Convention on Climate Change (UNFCCC), the UN Convention on Biological Diversity (CBD) and the Forest Principles, the latter being an illustration of the importance of deforestation as an environmental issue at the time.

Agenda 21 did, without doubt, galvanize opinion in the developed world concerning the environment. As we shall see later in the book, the institutionalisation of sustainable development within this document and its successors is probably the most significant shift in environmental policy ever witnessed. Yet *Agenda 21* is likely to be remembered more for starting a process than setting the targets and rules for ameliorating some of the problems it identified. In the first instance, the document is conspicuous in terms of the lack of targets that it sets. It is fundamentally aspirational in nature and to this end has been termed a 'political fix' by many. This is perhaps more evident within the context of developing nations. These countries had hoped that *Agenda 21* would mark a starting point for tackling the power and influence of trans-national corporations and the associated problems between economic inequalities created by their involvement in developing economies. Yet the wide sponsorship of the UNCED Conference by corporations meant that such concerns were never brought to the fore. Indeed, Connelly and Smith (2003) have argued that the Conference and resulting communiqué were couched in an ecological modernisation discourse, highlighting the role of technology in resolving environmental issues, rather than attempting to examine the underlying problems identified by developing nations. Overall then, many have viewed the UNCED as a talking shop and a way of perpetuating existing systems of economic development, rather than questioning the nature of these free-market approaches in depth.

However, as a footnote to these arguments, it is difficult to imagine a meeting with delegates from almost every nation being able to convene outside the context of the dominant free-market system that is the hallmark of most advanced societies today. What is remarkable, when seen in the context of current debates over climate change, is that so many did share an interest and did agree to the principles in *Agenda 21*. The acknowledgement of the importance of individuals to the process of change was vital and the impact institutionally could not have been imagined. For all of the criticisms that it is possible to level at Rio, it must be accepted that it was probably the most satisfactory political solution that could have been reached.

From the perspective of this book, Rio achieved something that no other international meeting had really attempted to do in the past. This can be illustrated by Principle 10 of the Rio Declaration, where the importance of participation was emphasised. This was a clear acceptance that environmental issues are dealt with, in many cases, most effectively by individuals and groups. The importance of environmental awareness was seen as vital in tackling many global environmental issues such as climate change. The responsibility for promoting such awareness was clearly delegated to national, but more commonly, local authorities. This reliance on local sustainable development, placed in the hands of individuals, was seen as essential if major global (and local) environmental issues were to be effectively tackled. The importance of individuals in the formation of European, UK and local policy will be further investigated in Chapter 3.

The Post-Rio Climate

The commitment shown at Rio has been continued by the United Nations. Most notably the Johannesburg summit of 2000 reviewed progress towards sustainable development and renewed pledges made in 1992. In Chapter 3 the political and policy responses to Rio will be examined in greater detail, but far more significant before we move to examine the nature of sustainable development, is to emphasise the cultural impact sustainable development has had, especially in the developed world. This can be viewed both positively and with some scepticism. The word 'sustainable' is currently one of the most popular (in quantity) words in our lexicon. As a discourse, or set of discourses, sustainability has become embedded into our everyday language. As a term, it proliferates a range of contexts, from when we visit the supermarket (sustainable food from sustainable agriculture), go to work (a sustainability officer to ensure that as an organisation there is sustainability), or go on holiday (sustainable tourism in the form of sustainable use of resources such as water or waste). To this end, we might conclude that sustainable development is already a huge success. Yet the proliferation of this term neither provides an assurance that everyone is clear about its definition nor does it imply that such messages translate into changes in attitudes and behaviour. For example, what does the term 'sustainable food' imply? This term is used frequently by retailers to brand specific products. Taking a cursory glance at Brundtland's definition, one might conclude that the product in question had been produced without causing damage to the environment, perhaps by using organic methods. Perhaps the product had been produced under socio-economic conditions that ensured no part of the production process was inequitable or prejudiced certain groups? Indeed, even if we were aware of the answers to these questions, would we know the full story in terms of the product's history and be able to make an informed judgement? These issues raise the question of how, if at all, it is possible to define sustainable development. We have already seen that the UNCED implicitly represented a particular perspective on this issue, highlighting the role of the free-market and growth as vital within the context of sustainability. However, to appreciate the complex and evolving nature of sustainability, it is necessary to examine how sustainability is framed at a series of levels. Such framing is based on differing principles, concepts and applications, and within the political discourse which as been described above, and in turn our response to such political developments will be framed by our own values and ethics.

Sustainable Development: Principles, Concepts and Application

The definition of sustainable development used by Brundtland (WCED 1987, 43) is the most widely known and often quoted definition of sustainable development: '…meeting the needs of the present without compromising the needs of future generations to meet their own needs'. Yet within this apparently simplistic statement lies a series of layers that need to be uncovered and examined before a clear assessment of sustainability can be undertaken. The guiding framework for such an examination of sustainable development is provided in Figure 2.1. This divides the study of sustainability into three distinct elements: principles, concepts and application. These will be dealt with in turn, but firstly the 'principles' enable us to understand and appreciate the diverse meanings and knowledges associated with sustainable development. Within this element, values refer to the underlying principles that guide the behaviour of individuals and frame the attitudes we hold towards a whole range of behaviours. In the case of the environment, a study of values enables the researcher to understand the attitudes of individuals towards specific policies. Ethics refer to the moral basis of the decisions taken. In environmental terms, ethics pertain to our perceptions of the value of life (human and non-human), equity between humans and societies and the basis of this equity, such as human needs. Finally, knowledges refer to the disciplinary contexts from which authors and policy-makers originate. The starkest example of the impact of disciplinary differences is inevitably between the natural and social sciences, but within these boundaries there are also significant variations.

Secondly, the concepts that underlie a study of sustainable development have already been touched upon in Chapter 1, but are significant with specific reference to sustainable development. Time, space and the notion of capital all play a significant role in our interpretation of both what can be seen as 'sustainable' (literally, durable and lasting) and what constitutes 'development' (in terms of scale and capital accumulation).

Within these two guiding frameworks, it is possible to move on to examine the key elements that relate to the 'application' of sustainable development, notably the dominant sectors, models and means of implementation. The three most commonly cited sectors of sustainable development are related to the environment, the economy and society. Each of these sectors needs deconstructing in terms of their definition and meaning. By doing so, it is possible to move forward to an examination of the (numerous) models that have been developed to understand and implement sustainable development. This text will merely scratch the surface of such models, looking in particular at three significant frameworks: the 'harmonisation' model (viewing environment, economy and society as equally proportioned), the 'nested' model (implying that environmental limits still set

VALUES	ETHICS	KNOWLEDGE
Social Environmental	Equity Needs	Disciplines Positionality

CONCEPTS

Time   Space   Capital

SUSTAINABLE DEVELOPMENT

SECTORS	IMPLEMENTATION	MODELS
Economy	The Policy Cycle	Harmonisation
Environment	Top - Down	Nested
Society	Bottom - Up	Weak / Strong
	Participatory mechanisms	

Figure 2.1   Key concepts for exploring sustainable development

the boundary around which there can be a sustainable society and thriving economy) and finally the 'weak / strong' model (adopting different interpretation of sustainability depending on the definition of capital applied).

Through the use of these models, we can examine the alternative possibilities for implementing sustainable development that are currently available. In the first instance, the nature of policy-making, in whatever form and at whichever scale, will be examined in the context of the policy-making cycle, emphasising the evaluative and iterative nature of policy formulation and reflection. On that basis, the text will move on to examine two classical models of policy-making along the spectrum of policy choices, the first reflecting a 'top-down' approach and the second 'bottom-up' approach. Building on this second perspective, the role of participatory mechanisms and public engagement will be examined as a preferred model for promoting sustainable development. This will provide the relevant context for both an examination of sustainable development policy and the role individual citizens have to play in the framing, analysis and resolution of environmental issues that lies at the heart of Chapters 4 to 9.

One significant point that needs to be addressed before this process of examination begins relates to the nature of research in sustainable development. Brundtland's (WCED 1987) report and later *Agenda 21* have provided researchers with such a wealth of theoretical and conceptual material to interpret that a reading of texts in this area would imply that there are multiple discourses surrounding sustainable development. As will become apparent when knowledges are examined, disciplinary perspectives are vital to appreciate when reading and interpreting work on sustainability. This text attempts to provide a constructive perspective on sustainability, drawn from the discipline of geography, whose implicit interest is the interaction of natural and social processes.

Principles for Sustainability

The principles that underlie a study of sustainable development relate to values, ethics and knowledges (Figure 2.1) and refer to the diverse backgrounds from which individuals are drawn to study sustainable development. As underlying and guiding principles in our daily lives (Schwartz 1977; 1992), *values* provide the basis for informing and making decisions on both the most mundane and most fundamental aspects of our lives. Values are driven by a range of social contexts, but are formed early in our lives. Schwartz (1992) has argued that although values are inherently personal in nature, there are certain 'universals' that can be traced within and between societies. In his study of over 100 countries, Schwartz argued that there were a set of social values that were common in these societies. This same argument has been applied to the environment by a wide range of authors, from both theoretical and empirical backgrounds. These studies are drawn from a range of social science disciplines, including economics, sociology, psychology and geography. The first set of values that can be established in any study of our attitudes towards environmental issues pertain to our relationship with the natural environment. Dunlap and Van Liere (1978) in a seminal study of values and environmental behaviour in the United States argued that 'biocentric' and 'anthropocentric' perspectives on the environment were significant in guiding the environmental action of individuals. These values should be seen on a continuum, with an extreme biocentric position representing values that position humans as equals with nature and makes no distinction on the basis of hierarchy between human and non-human. An extreme anthropocentric position holds that humans are distinctly differentiated from and, in hierarchical terms, 'above' nature. In reality, using such spectra inevitably means that each individual can be placed at a different point along the spectrum. For example, many individuals might be said to hold an anthropocentric position overall, believing humans to be distinct from nature, but would not necessarily view non-humans as 'inferior'.

This is where the importance of moving beyond such 'relational' values is important. What Barr (2003; 2004) has termed 'operational' environmental values enable us to understand more effectively how values can impact on the decisions that humans take towards environmental issues. O'Riordan (1976; 1985) has described a spectrum of values running from 'ecocentric' to 'technocentric'. Ecocentrists recognise the intrinsic value of the environment, viewing environmental resources as having value in their own right. An ecocentrist would also argue that there are clear and definable limits to human growth, which if exceeded would lead to irrevocable damage to the environment. This damage is viewed as negative from a perspective which argues that the natural environment is valuable in and of itself. At the other extreme, technocentrists view the environment as a resource to be exploited and that is extendible, so long as human interests are not compromised. From this perspective the environment has no intrinsic value, only economic value. Once again, each individual could be placed on this spectrum, with very few adhering to either extreme. For example, a mild technocentrist would acknowledge that the environment is valuable for human development and economic growth and would seek to use technology to

Environment and Society

improve efficiency, having some recognition that a specific level of natural capital stock is necessary for continued growth.

As Pepper (2003) has argued, there are numerous 'positionalities' relating to the environment that there is not space here to examine in detail. However, it is important to note the significance of appreciating the role that our underlying values have in framing our views towards the environment. In summary, how we relate to the environment and our views on the use of the natural environment for human development will have a major impact on the type of 'sustainable development' that we envisage and promote.

A further principle that we can examine relate to the *ethics* of sustainable development. Ethics relate to our treatment of other human and non-human actants and are partly a reflection of our values in terms of our attitudes towards how we value life and what position we take on issues such as equity. As Moffatt (1996) notes, the Brundtland Commission (WCED 1987) considered that all human and non-human life had value. Ethically, it is not clear how Brundtland differentiated the two in terms of the ethics of non-human life, in particular whether such life was intrinsically valuable or attainted value from an anthropocentric perspective. Indeed, as with environmental values, environmental ethics must be seen along a spectrum. For example, one extreme form of environmental ethics would argue for the fundamental rights of the environment and would view any exploitation of the environment as a violation of non-human rights. However, sustainable development has most often been related to our ability, as a generation, to provide an ethical basis for handing down the same level of resource as we owned. The key word here can be drawn from Brundtland's (WCED 1987, 43) own definition of sustainable development pertaining to the terms 'needs'. How 'need' is defined relates to our ethical position. The inequalities emphasised by Brundtland pointed to key problems in the distribution of wealth and resources in the global economic system (Turner 1993). Although we produce more food per head of population than ever before, this is unequally distributed, as is the wealth with which to purchase such commodities. Ethically, is it a sound position to argue that as a world community we should pass on *as little* to the majority of the next generation as we inherited? Indeed, is it ethically sound to pass fewer natural resources on to the next generation? Such questions inevitably highlight our sense of what is necessary for this and future generations to regard as 'quality of life'. The notion of 'need' has been examined in detail by Maslow (1978), who argued that over and above basic needs (such as safety, security and physiological requirements), individuals have a series of 'self-actualisation' needs relating to aspects such as love, truth, service, justice, perfection, aesthetics and meaningfulness. Brundtland implicitly highlighted many of these needs in the Commission's final report, yet one question that Maslow's (1968) now dated work does not address is the ethics of the consumer society that pervade western culture and have been blamed by some for exacerbating the poverty of those in developing nations. For many in contemporary western society, consumption and the re-casting of need in material terms poses significant problems for the ethical dimension of sustainable development. If it were ecologically possible to provide all six billion citizens on earth with standards of living commensurate with Europe or North America, this would be accepted by most. Yet this possibility seems unlikely given current ecological and technological constraints. Accordingly, we face the *ethical* dilemma that for western societies, sustainable development either means reduced growth or an acceptance of the status quo in terms of the wealth divide between nations. After Rio, many NGO's and organisations presenting interests in the South argued that the ethical position adopted by the North was not acceptable, citing the ethical flaws in maintaining trade and banking systems that exploited poorer nations (Adams 2003; Connelly and Smith 2003). Accordingly, the ethical basis for sustainable development is an important principle to appreciate, given the divisions and contrasts that it can reveal between the approaches formulated by both institutional and reflected by individuals in terms of their attitudes.

A final principle that needs to be considered is the *disciplinary* perspective from which both academics and practitioners derive their arguments. The range of authors that have written key texts in this area is testimony itself to the variety of disciplines with which sustainable development engages. Perhaps the most prolific work on sustainable development has been undertaken by economists such as Pearce (1991; 1993), Pearce et al. (1989) and Turner (1993) during early development of the concept. Crucially, the David Pearce School, later to become known as CSERGE (Centre for Social and Economic Research on the Global Environment) provided a series of *Blueprint* reports on developing a 'green economy'. The assumption made by these economists related to the importance of including the environment as both a resource, but crucially as a *cost* to be added into calculations. Thus, from an economic position, Pearce and his colleagues attempted to examine how environmental resources could be more effectively accounted for in the economic system and how the whole environmental cost to economic activity could be valued. Underlying this and arguments from a range of policy-makers was the assumption that technological innovation and modernisation could be used as a way of offsetting negative environmental consequences, whilst maintaining current growth trends. This technocentric position has become known as 'ecological modernisation', whereby advances in technology (such as energy saving devices and renewable technology) create efficiencies in the economy whilst providing environmental benefits. Therefore, taking an economic position on sustainability focuses on treating the environment as a resource which, if damaged, incurs a cost to those who have exploited that resource (known as 'polluter pays'). Reducing environmental damage can logically be achieved through increasing the costs of exploitation and damage through economic penalties. Such penalties are not restricted to the primary users of such resources, but are passed on to individual consumers as higher costs and disincentives. The adoption of less damaging practices is, by contrast, promoted by the use of incentives to encourage a shift in behaviour.

This economic perspective implies that exploitative behaviours (of organisations and individuals) can be changed through focussing on costs and benefits, with a clear notion that higher costs will lead to a change in attitudes and behaviour. Yet a cursory glance at other social science disciplines provides evidence that this purely economic approach is not given the same level of prominence when the issue of individual behaviour change is considered. Psychologists, for example (see Oskamp 2000, for an excellent review), have focused on the role that individuals can play in reducing their impact on the environment, without necessarily taking into account economic penalties (or incentives). Considerable psychological research suggests that environmental concern amongst individuals is high across western societies and the critical problem is not one related to convincing individuals of the *need* to change their behaviour, but rather a series of complex structural and psychological changes that are required to enable individuals to realise these environmental concerns.

From a more ecocentric position, environmental sociologists such as Dunlap et al. (2000) have studied the potential for a change in social values towards the environment and consumption. The clear assumption here relates to the potential for individuals to adopt lifestyles that are less consumptive and materialistic. To this end, such research reflects the work of a range of geographers (e.g. Barr 2004; Burgess et al. 1998; Burton 2004; Eden 1993; 1998; Wilson 1997) who have studied sustainable development in terms of the changing attitudes and behaviours of the public towards the environment, again basing their assumptions on a socially-constructed model of nature-society relations. Unlike the economic interpretation of sustainable development, researchers from other social sciences therefore view sustainable development as a social *process*, which does not necessarily lead to an end point.

These perspectives stretch across social science disciplines, yet the interpretation of sustainability is once again contested when examined from a natural science perspective. Within this context, the environmental aspect of sustainability attains greater status, with economic and social systems relying on and operating within environmental limits. Accordingly, one of the most problematic areas of current sustainable development policy lies in the inability of researchers from the natural and social sciences to engage in research that attempts to bridge this gap.

Concepts for Understanding Sustainable Development

Just as principles for sustainable development enable an understanding of the backgrounds which individuals bring to sustainability (values, ethics and knowledges), the concepts underlying sustainable development provide the lenses through which the concepts can be viewed. As Figure 2.1 illustrates, the concepts of time, space and capital enable us to take a closer look at how sustainable development is framed in different contexts. Of these, time is perhaps the most transparent element of 'sustainable' development, emphasising the notion of futurity (Baker 2004), with Lele (1991), noting that the term 'sustainability' simply refers to durability. Turner (1993) develops this notion by referring to the intergenerational nature of sustainable development, which as we have seen from an ethical perspective, necessitates providing future generations with the same level of opportunities and resources that current generations experience. Apart from the ethical connotations of these issues (see Moffatt 1996 and Turner 1993) this notion of sustainability in an intergenerational context presents specific problems to both policy-makers and ourselves as individuals. As Meadows et al. (1972)

highlighted (see Figure 1.1), our human perspective to consider how our own lifestyles and activities will impact on future generations is highly problematic.

This temporal problem is also reflected by spatial concerns. Since the *global* concerns surrounding sustainable development have emerged, the emphasis on the global dimension of sustainable development has slowly been replaced by local interpretations and policies. This addresses Meadows et al.'s (1972) second concern, that our ability to appreciate impacts on the natural environment reduces when we move beyond the local scale. Nonetheless, messages related to sustainable development contain a range of spatial metaphors, which need to be borne in mind when considering the concept. Redclift (1991) highlights two significant elements to this spatial dimension. First, he notes the importance of 'levels' of sustainability, citing differing examples of scale from (for example) the farm scale, the field, the village and so on. Redclift (1991) points out that sustainability at one scale may not imply sustainability at another scale. A good example is the management of household waste. Locally, a suitable level of provision may be available for the land filling of waste. This localised spatial consideration does not necessitate a shift to recycling or waste reduction. However, this may be at odds with the global 'problem' of greenhouse gas emissions, a major contributor being methane gas from landfill sites.

Indeed, as Redclift (1991) points out, the spatial dimension also raises political concerns. Definitions of sustainability will vary depending on the specific political and cultural context, most evidently epitomised by alternative perspectives of sustainable development between the North and the South (Adams 2003; WCED 1987; Pearce 1993). In a Northern context, sustainability is often framed in terms of the sustainability of our current development and how, at the very least, to maintain that development and quality of life. In the South, discourses may pertain more to the potential for development and growth to enhance quality of life, with less emphasis being placed on Northern interpretations of sustainable resource use.

Finally, there is the notion of capital. In an effort to appreciate the nature and volume of different resources, economists have used the notion of natural capital as a way of valuing the environment (Pearce 1993). Pearce's (1988) perspective on capital can assist us in appreciating the basis for different models of sustainable development. Pearce argued that natural capital could be divided into three categories:

- critical natural capital: this is capital that is required for survival. it can be viewed as either functional (such as the presence of the ozone layer or the atmosphere in general) or valued (a good example being rare species, valued in terms of their potential for health care);
- constant natural capital: this is capital which must be maintained in some form, but can be adapted or replaced (a good example being the adaptation of natural woodland for a natural park);
- tradable natural capital: this is natural capital which is not scarce or highly valued and which can be replaced.

Economists have used these notions of capital to understand how environmental resources can be more effectively utilised in the economic system. The importance of critical natural capital, however this is defined, is significant because it is regarded as essential for life to continue. Assigning resources as critical, constant or tradable natural capital is, of course, a highly contested subject and one which those holding alternative ethical positions will debate. However, the notion of capital is the means by which researchers and policy-makers value not just the environment but all aspects of sustainable development.

Applications of Sustainable Development

Both the principles and concepts of sustainability provide the lenses through which to view a study of sustainable development; they enable us to appreciate the backgrounds from which individuals are drawn and the perspective they take on sustainability. However, there are also a variety of approaches to applying sustainable development, as Figure 2.1 indicates. Until this point, we have dealt mainly with environmental and, to a lesser extent, economic issues. Yet sustainable development in its simplest form relates to an interest in the environmental, economic and social dimensions of human life, at all temporal and spatial scales. Khan (1995) provides a good overview of what he terms this 'paradigm of sustainable development' (see also definitions of these concepts in Baker 2004 and Turner 1993). Figure 2.2 presents this paradigm in diagrammatic form. What Khan perceives as the central elements of each component of sustainable development is given in each segment of the circle. Beginning with environmental aspects, Khan states that (1995, 65):

> Environmental sustainability ... means that natural capital must be maintained, both as provider ('sources') or inputs and as 'sink' for wastes.

In both instances, Khan argues that regeneration rates need to be ensured, such that the environment's ability to cope with waste is not irrevocably overwhelmed, nor indeed its ability to regenerate source materials overcome. Economic sustainability is, according to Khan (1995, 64) '... a production process that satisfies the current level of consumption without compromising future needs'. Khan points out that the *modus operandi* of this rather vague notion has changed in recent years. Previously, economic sustainability had relied on the assumption that the market would encourage innovations in technology required to become more efficient, slowing the rates of reduction in critical natural capital. The realisation that this was not occurring has led many to argue that economic sustainability needs to account for the relative slow pace of technological development in contrast to the rapid pace of environmental resource exploitation (Khan 1995). Finally, social sustainability emphasises the need to alleviate poverty within a framework that promotes equity, empowerment and cultural identify. These developments are seen as vital in producing both economic and environmental sustainability. The links clearly demonstrated by Brundtland (WCED 1987) between poverty, lack of development and environmental issues, have forced

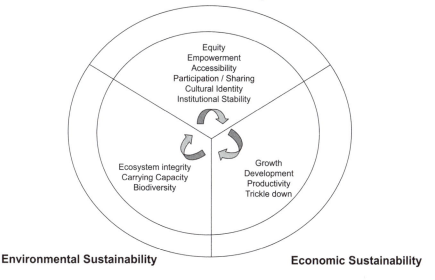

Social Sustainability

Equity
Empowerment
Accessibility
Participation / Sharing
Cultural Identity
Institutional Stability

Ecosystem integrity
Carrying Capacity
Biodiversity

Growth
Development
Productivity
Trickle down

Environmental Sustainability   **Economic Sustainability**

Figure 2.2   Khan's (1995) paradigm of sustainable development. © John Wiley and Sons Limited. Reproduced with permission

Source: Khan (1995).

social sustainability up the political agenda (Khan 1995). Nonetheless, although originally focused within a developing world context, social sustainability is seen as universal in promoting the basic value shown in Figure 2.2. It is interesting to note that the social sustainability agenda, whilst containing the most key words and in many ways, the most desirable goals, presents a series of aspirations that are the most problematic to measure.

Outside of this basic framework, other researchers have sought to add to or alter these three common themes. Although they provide useful insights, the dominant popular discourse surrounding sustainable development undoubtedly uses the three concepts of environment, economy and society. However, one of those who has sought to re-frame the debate is Basiago (1995) who has argued that sustainability as a construct operates within alternative dimensions (similar to the discussion above relating to knowledges). To this end, Basiago argued that to understand sustainable development, alternative lenses had to be used to view sustainability. He identifies four such lenses:

- biological sustainability: an argument that views sustainable development being concerned mainly with biodiversity;
- economic sustainability: seen in five separate contexts:
  - environmental: relating to the physical aspects of the earth and solar system (lithosphere, biosphere, atmosphere and the sun);

- o human: related to aspects such as health, knowledge, skills and motivations of people;
- o socio-organisational capital: our habits, norms, tradition; law, government;
- o manufactured: including buildings, equipment, goods;
- o credit: all money and debt;
- sociological: sustainability as social and environmental justice:
  - o concerned with land and population contamination;
- planning:
  - o a concern for future generations;
  - o a concern for the future global environment;
  - o a 'joined-up' structure;
  - o a 'new pedestrian' society.

Basiago's alternative construction of sustainable development is useful because it represents many of the implicit assumptions made by economists regarding the environmental value of resources. Yet one area that Khan (1995) would emphasise is explored by Franks (1996). Although social, economic and environmental constructs are important to delivering sustainable development, the institutional aspects must also be considered. Franks (1996) argued that related to the notion of social sustainability, major changes were required in how organisations (for example governments, NGOs, businesses) functioned and viewed their role in terms of sustainable development. The significance of institutional sustainability is a movement away from a narrow definition of the institution's specific interest towards a consideration of all interests. In essence, Franks is referring to what we would colloquially term a 'culture shift'.

Such a shift in values, ethics and cultural assumptions can help us to answer the crucial question which needs to be addressed before moving on: what makes *sustainable* development any different from other forms of development? Turner (1993, 5) provides a useful starting point 'Sustainable development ... is ... economic development that ensures over the long run'. Within this simple definition, Turner highlights the role of wider definitions of 'economic', to include education, health and quality of life. Continuing the institutional theme, Khan (1995) has argued that sustainable forms of development have three major underpinnings:

- institutional: an understanding that development must be seen within natural limits and be sensitive to intergenerational issues. Indeed, institutionally, all social costs and benefits must be accounted for;
- cultural: a movement away from the dominant social paradigm of consumption and materialism;
- international relations: a clear regulatory framework for development.

This represents an inspirational set of differences between development and sustainable development. In reality, all that can be said is that sustainability at present does acknowledge the role of environmental protection, economic development and social progress, but the sharp differences in how sustainability is interpreted means that pointing to a fundamental shift in individual attitudes towards development and growth is yet to materialise and is unlikely to at present.

Models of Sustainable Development

The preceding applications of sustainability succinctly describe a set of 'sustainabilities'; a way of imagining sustainability (as either an end-state or process). Yet a number of authors have attempted to move beyond these more conceptual approaches to explore the relationships between different 'sectors' of sustainability (Figure 2.1). At a simply level, the description of sustainable development provided by Khan (1995) makes reference to the relationships between the economy, environment and society. Figure 2.2 emphasises that equal priority is given to each of the three segments, with no assumptions being made regarding the limits that one may place on another. This 'harmonisation' model is a popular representation of sustainable development, one which Giddings et al. (1996) referred to as the 'three rings' model of sustainable development (Figure 2.3). Conceptually, this framework does provide a means by which to classify and represent the different interests in each element, yet in reality the conflicts between the three segments necessitate a more sophisticated approach to examining the relationships between these three scenarios. For example, Khan (1995) acknowledges the crucial role of environmental limits in regulating economic activity, yet his framework does not offer a means by which to characterise this relationship. Accordingly, it is necessary to examine how researchers have sought to assess the role of the environment, economy and society and how we can begin to recognise and classify alternative forms of sustainable development within different economic, social and political contexts.

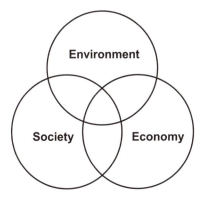

**Figure 2.3   The 'three rings' model of sustainability (after Giddings et al. 2002).
© John Wiley and Sons Limited. Reproduced with permission**

Source: Giddings et al. (2002).

Giddings et al. (1996) provided a detailed critique of the 'harmonisation' model of sustainable development. They argued that the notion that the three 'rings' of the environment, economy and society can be classified as separate entities but brought together to create a balanced approach to satisfying the requirements of each is misleading. Part of the difficulty relates to the 'symmetrical interconnection' (189) of the relationship between the three sectors. Giddings et al. question why this should be the case, given that acknowledging the difference between sectors implies that there is likely to be prioritisation of sectors. Building on this argument, the authors highlight the weaknesses inherent in the 'three rings' model:

- it tends to encourage a compartmentalised view of sustainability, taking little account of the links between the three sectors;
- in an attempt to maintain the inherent balance (see Figure 2.3), trade-offs are encouraged between the sectors, which may include all types of natural and other capital. This encourages a 'technical fix' approach to sustainability;
- it encourages a compartmentalised view of the environment as an 'other' place, such as wilderness, separating it from the everyday consciousness of society and the economy.

Giddings et al. (1996, 190) presented two alternatives to this approach. In the first, which they term the 'political reality' they argued that 'The reality of life today is that the economy dominates environment and society'. This political reality is reflected in both popular political discourse (the now infamous comment by Bill Clinton in his 1992 election campaign for President of the United States that 'It's the economy stupid' is illustrative of this point) and wider academic literature. The dominance of the early sustainability debates by environmental economists has encouraged the use of terms such as 'capital' as a means of framing sustainability. Accordingly, it is without doubt that economic priorities are normally the ones highest on the political agenda at election times and, to turn the issue on its head, it is extremely unlikely that any government in the UK would ever be removed from office for environmental mismanagement, in the way that government's have been removed in the past for their poor stewardship of the economy.

Political reality is, as Giddings et al. (1996) suggested, just that, a tool by which to lever power within constitutional democracies. Accordingly, they term their second alternative the 'material reality':

Society embraces the multitude of human actions and interactions that make up human life. Without society, humans would not survive, as our very existence, both evolutionary and present terms, is based on social interaction. Human activity takes place within the environment. (Giddings et al. 1996, 191)

This led the authors to propose an alternative model of sustainability termed 'nested sustainability' (Figure 2.4). The essence of this model is to acknowledge human reliance on nature and our bounded presence within it. A two-way process is conceived, in terms of both limits to environmental exploitation and also the impact that human activities have on the environment. The economy is placed within a societal framework, given the reliance of economic activity on social exchange and networks. Indeed, Giddings et al. argued that many human needs are met outside what would traditionally be defined as 'the economy'. Accordingly, the nested model of sustainable development attempts to provide a hierarchical structure to the three rings model, specifying the nature of the relationships between the three sectors and their relative importance.

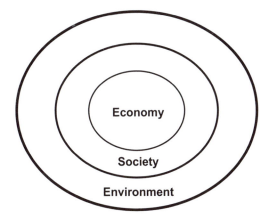

Figure 2.4 Giddings et al.'s (2002) nested sustainability model. © John Wiley and Sons Limited. Reproduced with permission

Source: Giddings et al. (2002).

In interpreting this model of sustainability, we can begin to see the value of studying a set of principles as a way of interpreting such a framework (Figure 2.1). In terms of the values and ethics that this model characterises, we could argue that it presents an approach grounded in an ecocentric view of society – nature relations, by placing the emphasis on environmental limits. Ethically, the model places a greater emphasis on social needs rather than economic priorities. Epistemologically, the knowledge base from which this model originates is reflective of a sociological interpretation, rejecting the arguments of environmental economists, who have taken a technocentric approach towards sustainable development. Likewise, by using the concepts of time, space and capital, we can critically examine this model. Without doubt, such a framework is capable of being used over a range of both temporal and spatial scales, but most significantly, it rejects an economic definition of capital, preferring instead to emphasise the value of each component and implicitly viewing trade-offs as negative policy instruments.

These two models of sustainable development present radically different approaches to conceptualising what Brundtland (WCED 1987) had defined in her Commission's report. As we have seen, these differences are based on principles (values, ethics and knowledge) as well as concepts (time, space and capital), alongside the related debates of how the economy, environment and society are related to each other. However, models as interpretive tools have limitations, not least inflexibility: they present one way of conceptualising an issue or problem. Accordingly, researchers have sought alternative and more sophisticated ways of appreciating the range of assumptions that are included in sustainability policies. This has led to the development of what is widely known as 'weak' and 'strong' sustainability approaches.

Williams and Millington (2004) have termed these approaches as 'shallow environmentalism' (weak sustainability) and 'deep ecology' (strong sustainability). The terms were originally developed by environmental economists such as Pearce (1993) and Turner (1993) although they have been adopted and adapted by other social scientists (e.g. Baker 2004; Gibbs et al. 1998). At the heart of this approach towards sustainable development is the distribution of capital. A strong approach towards sustainable development has at its heart the desire to maintain critical levels of natural capital, defined by Pearce (1988) as either functional or valued (see above). A weak approach to sustainable development argues that natural capital, of any type, can be traded-off and substituted with human capital, so long as the total capital passed on to the next generation is constant or growing. The diagram in Figure 2.5 illustrates this process of intergenerational capital exchange for these two approaches.

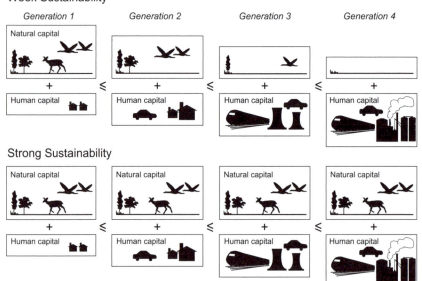

Figure 2.5   Weak and strong approaches to sustainable development (after Roberts 2004)

Source: Roberts (2004).

Although Turner (1993) provides an account based on environmental economics, Pearce (1993) developed a continuum which he used to position weak and strong forms of sustainable development in terms of both environmental and political values and ethics. An adapted version of this is provided in Figure 2.6, with additions from Baker (2004) and the current author. The first point to note is that each approach is divided into two, providing both, for example, 'very strong sustainability' and 'strong sustainability'. Second, examination of the table renders it possible to explore the potential ethics, values and concepts associated with each approach. An interesting question at this point might therefore be (as an exercise in interpretation) where one would place the UK on such a spectrum? It could be argued that the UK can be located at the 'very weak' end of the spectrum, given the reliance the UK economy places on free market principles and a general absence of regulation.

In this section of the chapter we have examined three ways of examining sustainability. There are many others, but these three ways of approaching sustainable development represent many of the more detailed models that are available. What these approaches have demonstrated is the complexity of both theorising and practicing what appears at face value to be a very simple concept. As we have seen, these models and approaches are representative of a series of layers that need to be uncovered to understand how different researchers derive their operational frameworks.

Implementation

A final element noted by both Brundtland (WCED 1987) and included in Agenda 21 (UNCED 1992) was the importance of implementation (Figure 2.1). This needed to involve several key elements: regulation (in the form of legislation), government and institutional action, participation and evaluation. Traditional approaches to policy development until the 1990s were dominated by what we can term a 'top-down' approach to policy (Gilg 2005). This form of policy-making assumes that governments, as honest brokers and experts on their field, have the relevant knowledge and experience to formulate and implement policy from a centralised perspective. *Agenda 21*, although not legally binding, did include significant agreements from national governments on their role in providing leadership for sustainable development. Yet Brundtland's (WCED 1987) report clearly emphasised that such a 'top-down' approach had to be seen within the context of the many problems posed by a centrally-managed policy structure. Such structures had led to catastrophic policy decisions in the past, such as mass settlement of deforested land in South America. The importance, therefore, of a range of stakeholders was vital to policy formulation, requiring consultation and democratic accountability.

This alternative form of policy-making, termed the 'bottom' up' or 'grassroots' approach emphasised participation in decision making from all stakeholder groups (Baker 2004). The 'participatory' agenda was the hallmark of sustainable development policy in the 1990s and reacted to Brundtland's call for key groups in society (such as women, NGOs and indigenous peoples) to be involved in

	TECHNOCENTRIC		ECOCENTRIC	
	Cornucopian	*Accommodating*	*Communalist*	*Deep ecology*
GREEN LABELS	Resource exploitative, growth-orientated position	Resource conservationist and 'managerial' position	Resource preservationist position	Extreme preservationist position
TYPE OF ECONOMY	Anti-green economy, unfettered free markets	Green economy, green markets guided by economic incentive instruments [EIs] (eg pollution charges etc.)	Deep green economy, steady-state economy regulated by macro-environmental standards and supplemented by EIs Zero economic growth; zero population growth	Very deep green economy, heavily regulated to minimise 'resource-take' Reduced scale of economy and population
MANAGEMENT STRATEGIES	Primary economic policy objective, maximise economic growth (Gross National Product [GNP]) Taken as axiomatic that unfettered free markets in conjunction with technical progress will ensure infinite substitution possibilities capable of mitigating all 'scarcity/limits' constraints (environmental sources and sinks)	Modified economic growth (adjusted green accounting to measure GNP) Decoupling important but infinite substitution rejected. Sustainability rules: constant capital rule	Decoupling plus no increase in scale. 'Systems' perspective-'health' of whole ecosystems very important; Gaia hypothesis and implications	Scale reduction imperative; at the extreme for some there is a literal interpretation of Gaia as a personalised agent to which moral obligations are owed
ETHICS	Support for traditional ethical reasoning: rights and interests of contemporary individual humans; instrumental value (i.e. of recognised value to humans) in nature	Extension of ethical reasoning: 'caring for others' motive - intragenerational and intergenerational equity (i.e. contemporary poor and future people) ; instrumental value in nature	Further extension of ethical reasoning: interests of the collective take precedence over those of the individual; primary value of component functions and services	Acceptance of bioethics (i.e. moral rights/ interests conferred on all non-human species and even the abiotic parts of the environment); intrinsic value in nature (i.e. valuable in its own right regardless of human experience)
SUSTAINABILITY LABLES	Very weak sustainability	**Weak sustainability**	**Strong sustainability**	Very strong sustainability

Figure 2.6   The sustainability spectrum (after Pearce 1993). © Earthscan. Adapted with permission

Source: Pearce (1993).

policy formulation. Baker (2004, 41) provides an examination of the value of participation for sustainable development policy-making. Normatively, the changes that society may need to make in order to move towards sustainable development imply that 'It is only through increased participation that society can construct a "shared public basis" on which to ground the legitimacy and acceptance of such restrictions and corrections'. Functionally, Baker argues that out of necessity, the all-embracing nature of sustainable development means that many groups in society need to be consulted to address their needs and concerns. Although this has formidable political obstacles, not least in attempting to represent alternative perspectives, a move towards achieving consensus has become one that is clearly desirable.

The nature of participation is, however, more problematic than the establishment of the principle that it is a social 'good'. Brundtland's (WCED 1987) assumptions underlying her call for greater participation were grounded in a belief that citizens wanted to be consulted and wished to participate. Yet the move from 'passive' to 'deliberative' participation has been slower than many had hoped. The analysis of why this has been the case is the subject for another book, but suffice it to state that most western societies have become de-politicised in the past 30 years and have generally witnessed reductions in passive forms of participation (such as voting). This democratic and political scepticism has therefore been even more problematic to overcome in terms of promoting deliberative forms of engagement. This is particularly problematic with a subject such as sustainable development, given the ambiguity of the concept and a tangible sense that the public are lacking in response efficacy.

It seems that the jury is still out in relation to deliberative forms of participation, yet the word 'participation' does not merely refer to decision-making, but a host of other activities that are relevant to sustainable development. This text partly aims to provide an alternative to the deliberative argument of participation, arguing that in many cases, citizens are participating in a range of activities to promote sustainable development. This is framed within the current political context and seeks to work towards a constructivist perspective on participation in seeking to explore new ways to encourage action for sustainability.

Conclusion

Almost 40 years after the so-called 'Environmental Crisis', the debate still rages over the relationship and significance of the environment and economy. If the content of debates has altered little, the tone of them certainly has. In the early 1970s, 'environmentalism' was a highly politicised, pro-active and in many cases 'radical' threat to the political establishment. It predicted global catastrophe and recommended radical economic reductionism to prevent population collapse. Although many of the core arguments remain the same, the onset of sustainable development and the mainstreaming of 'the environment' as a concept and policy issue has created both opportunities and problems for members of the environmental lobby. The chair of the UK Sustainable Development Commission,

Jonathon Porritt, said after Rio that 'I had low expectations of this Conference and all of them have been met'. Yet Porritt now heads the UK government's 'critical friend' on sustainable development, the Sustainable Development Commission. Some have therefore argued that environmentalism has lost its edge, giving in to a mainstream and fundamentally technocentric perspective. Others see the onset of sustainability as an opportunity to change policy for the better, within the system. Perhaps the question that emerges from this situation is whether, after the flurry of activity from Rio and the second Earth Summit in Johannesburg, the discourse of sustainability will last. Currently in vogue, sustainability is unlikely to outlast this generation without a radical overhaul. This is because, despite the current political climate, the enduring concern and perpetual conflict that emerges is between the environment and economy.

Yet despite these musings over the concept's future, it is abundantly clear from both the chronological and thematic material presented in this chapter that sustainable development has brought the citizens 'closer' to the environment. Politically, the *Brundtland Report* and *Agenda 21* emphasised the significance of 'bottom-up' approaches to resolving environmental dilemmas and the importance of key stakeholders, including 'the citizen'. Conceptually, this has been reflected in the debates surrounding sustainable development, away from a macro-level discussion on the conflicts between environmental and economic imperatives, towards a more sophisticated (some might argue blurred) approach, viewing sustainability as a concept which needs to both involve society and individuals in environmental protection and which seeks to improve quality of life through sustainable economic development.

However, although we have been thus far explored the notion of sustainability and the emergence of individual citizens and society as key actants, there is a need to situate this broad trend within the context of a specific political framework for the United Kingdom. Such a contextual underpinning will enable the reader to appreciate the scales at which sustainability policy in the UK acts to formulate the political framework for promoting citizen engagement in sustainability. Accordingly, the next chapter examines how the principles of sustainability have been institutionalised at a range of administrative scales. Beginning with the European Union (EU), the chapter examines the role of sustainability in moving EU policy from one concerned with pollution control to a sustainable development. The chapter then considers the response of the UK to the Brundtland Report, including the Conservative Government's Sustainable Development Strategy (DoE 1994) and the shifts in emphasis witnessed as Labour came to power in 1997. This is framed within an expanding social concerns and quality of life debates, away from core environmental issues. Within the UK, the chapter then examines the local context and implementation of sustainable development during the 1990s, with reference to the onset and success of Local Agenda 21 (LA21). This will be examined through the lens of participation, particularly focused around a critical review of LA21 policies for participatory engagement. The work of researchers such as Selman will be examined, which provides key insights into the role of LA21 in encouraging greater public participation. Finally, the chapter will consider Labour's moves to revitalise public participation at the local level and will consider the Community 2020 initiative and the shift to Community Strategies within local planning systems.

Chapter 3

Policy

We all, wherever we live, face a future that is less certain and less secure than we in the UK have enjoyed over the past fifty years. (DEFRA 2005, 12)

The publication of the 2005 United Kingdom Sustainable Development Strategy (DEFRA 2005) from which this emotive quotation above is taken, represents the most advanced and sophisticated perspective taken by the Untied Kingdom government on sustainability since the 1992 Rio Earth Summit. As we shall see later in this chapter, the emphasis placed on the all-encompassing nature of sustainable development, both in physical and social terms, was a position that was reached only after two previous strategies. Such a position represents the link in a policy-making process that has seen sustainability lie at the heart of a complex hierarchical structure, spanning international, European, national, regional and local administration.

This chapter will examine the nature of these structures, with specific reference to how discourses of participatory mechanisms for sustainable development have emerged since 1992. As will be seen in the forthcoming pages, the participatory agenda presents a narrative that has evolved from a growing dissatisfaction with the technocratic approach popular in the 1980s. The chapter will examine how policy has begun to shift from a 'top-down' approach to a 'bottom-up' approach. This problematises the relationship between the environment, the state and citizens, enabling voices that have previously been regarded as 'lay' and 'silent' to have efficacy in both framing and tackling environmental problems. To this end, British policy towards citizens and sustainable development has begun to reflect the complex and blurred relationships that reflect society-nature relations. In particular, there has been recognition of the implicit contradictions in heading towards 'sustainable lifestyles'. Once again, such contradictions can be conceptualised in terms of temporal and spatial dimensions. Temporally, the efficacy of recycling waste has short-term benefits to both the self and the wider environment. Cutting car use to reduce the impact of climate change does not produce the same levels of efficacy: such a decision appears to both reduce short-term benefit to the self in terms of convenience, as well as the short-term benefit of environmental protection. Similarly, in terms of space, recycling has tangible local environmental benefits, whilst reducing climate change relies on an intangible benefit to the world.

Accordingly, this chapter begins with an examination of the policy-making process and structures for delivering sustainable development. It then moves on to examine the nature of four major policy structures for sustainability: European, national, regional and local. Through the use of this approach, dominant themes in the citizen engagement debate will be introduced, culminating in an evaluation of policies for behaviour change.

Sustainability: Structures for Change

Sustainability, as a concept, has been integrated into existing policy structures with more success than any other policy initiative in the twentieth century. For example, *Planning Policy Statement 1*, originally *Planning Policy Guidance Note 1: Planning Principles*, is prefaced with the statement that 'Planning exists to deliver sustainable development' (ODPM 2005, 1). Such a sweeping statement is vital to our understanding of the role that sustainability has come to play in the everyday workings of government. From 2005, for example, all government departments have been briefed to 'proof' their respective policies for sustainable development.

   The drivers for these changes emanate both from national interests in sustainable development, but also the hierarchical nature of sustainable development policy. Instituted in 1992 at the Rio Earth Summit, Agenda 21 presented a series of work plans for government at various levels. Figure 3.1 provides the essential guide to these structures within a UK context with indicative examples provided in italics. The 1992 Rio Earth Summit set the policy agenda through the final Communiqué (UNCED 1992) in the form of *Agenda 21* (Chapter 2). Whilst the Rio summit was being prepared, the European Community (now European Union or EU) set about adapting its Environmental Action Programmes (see below) into a sustainability agenda. The complexities of European integration mean that a wide range of environmental legislation is passed by the European Commission to be implemented across the community. European environmental policy and legislation acts in two ways in terms of member state influence. In the first instance, European policy can directly influence the ability of a member state to enact legislation relating to a specific environmental issue. For example, much of Britain's waste policy is driven by the 1999 European Union Landfill Directive (1999/31/EC). The major impact of this directive was that member states had to reduce the biodegradable waste being sent to landfill to 35 per cent of 1990 levels by a specified date (2015 in the case of the UK). This in turn led to the publication of DETR's (1999) *Limiting Landfill* document, eventually culminating in a new national *Waste Strategy* (DETR 2000). However, European policy and legislation also acts to re-frame environmental and sustainability debates at the national level by providing a strategic direction for national policies. Part of this is inevitably 'second guessing' the European agenda, but mostly it is a question of dissemination of principles agreed at European level and translating them into a specific national context.

   Nationally, the institutionalisation of sustainable development policy has come in the form of two agendas: sustainable development policies and 'proofing' for sustainability throughout government. The former has inevitably preceded the later. In Britain, sustainable development strategy has been the preserve of the Department of the Environment (DoE), becoming the Department of the

Figure 3.1 Policy structures for sustainable development

Environment, Transport and the Regions (DETR) when Labour took power in 1997. This large department was subsequently disbanded and the strategic focus for sustainability policy is now the responsibility of the Department for Environment, Food and Rural Affairs (DEFRA). As sustainable development has attained greater prevalence, the dissemination of sustainability messages has infiltrated government departments, who have developed various strategies to address the issue.

The onset of regional administration in the UK since 1997 has also been influential in promoting sustainable development. Overall strategic responsibility for sustainability lies with unelected Regional Assemblies (RAs), who coordinate regional sustainable development strategies within a bewildering framework of policy structures, incorporating everything from regional health strategies to transport policy. Such complexity has been introduced since the system for plan-making in England and Wales was changed in 2005 to reduce the role of county council's in coordinating policy, in place of the regional scale. Central to this plan-making process is the development of the regional spatial strategy, within the context of the regional sustainable development strategy and all other relevant strategies.

Finally, local authorities have responsibility for developing sustainable development strategy insofar as the different constituent elements of sustainability fall into their remit. This can be a wide range of elements, including waste management, economic development and social inclusion. However, the more formalised nature of sustainability policy in local authorities has been focused around the development of what were known as *Local Agenda 21* plans, succeeded by *Community Strategies* within the context of *Local Development Frameworks*. These are community-led and locally focused strategies that have sought to increase participation amongst citizens for sustainable development. As will be seen later in this chapter, such strategies have witnessed limited success, often suffering from a lack of focus and institutional opposition to their implementation.

These four layers of government represent a formalised structure for implementing sustainability. Yet it would be misleading to represent this framework as an inflexible or dominant force in implementing sustainability. At every level of government, interests relating to various aspects of sustainable development are represented by a series of stakeholder groups. In some cases, non-governmental organisations (NGOs) have taken a leading role in influencing and implementing sustainable development. At the local level, this can mean radical departures from previous practice. For example, many local authorities have tendered-out the process of devising and delivering a Local Agenda 21 plan. Nationally, the government has actively promoted the role of key stakeholder organisations to feed into its policy-making forums.

Such a blurring of the lines of policy demarcation necessitates a brief discussion of how sustainability is developed as a policy response to environmental, social and economic development. Within the British state, there has been a tradition of developing policy with the views of key interest groups in mind. Figure 3.2 illustrates the role of such groups alongside the other major influences on sustainability decision-making. Figure 3.2 is of course simplistic and does not seek to provide a comprehensive analysis of the influence on decision-makers; rather it provides an illustration of the spheres of influence in which policy is formulated. Yet an often neglected component of the British political system must be recognised relating to the role of elected politicians on the one hand and civil servants on the other (the outer and inner circles respectively in Figure 3.2). Although the two are rarely seen to be in conflict openly, a political imperative has to be moulded by the ability of the civil service to implement a particular policy. Indeed, these conflicts are often illustrated most effectively at the local level, where decisions relating to developments are made against civil service advice. Accordingly, as Figure 3.2 demonstrates, the policy-making process is a constant negotiation between interest groups, politicians and civil servants. Having outlined these key principles of sustainability policy, we now move to examine each policy-making structure in turn.

The European Perspective

Despite the current scepticism both in Britain and the European Union more widely, 'Europe' has driven an agenda for change, especially in environmental policy, that has undoubtedly resulted in a higher level of environmental protection and conservation than would have been possible for individual member states

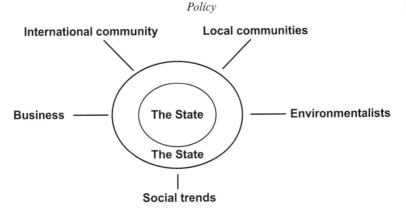

Figure 3.2   Conflicts and sustainability

to achieve in any other way. The popular image of European Union policy is perhaps not as measured as this and for many in the popular media, 'Europe' is viewed as a bureaucratic empire, seeking to extend power over member states. This book does not seek to join this debate directly, but a closer examination of European policy on the environment does reveal a desire (if not always reflected in reality) for an engagement of individual citizens in promoting its environmental agenda (see below). Yet most European policy-makers acknowledge that the scale at which 'Europe' operates means that environmental legislation can only set the boundaries for what member states choose to implement. Accordingly, Europe has been and remains the driving force behind considerable member state legislation on the environment, but its direct impact on citizens remains limited, not helped in Britain by the potentially hostile reaction to 'Brussels'. For more information about EU environmental policy, texts such as Barnes and Barnes (1999) or Jordan (2002) provide excellent introductions. This brief section deals with sustainability policy only.

Until 1992, European Union action towards the environment focused on a reactionary and 'top-down' approach towards policy-making. Four environmental action plans, begun in 1973, were designed to form the basis for introducing environmental legislation focused around specific areas of concern. However, the onset of sustainable development brought about a change in European environmental policy, re-orientating the focus towards a wider notion of sustainable development and taking account of Brundtland's (WCED 1987) recommendations relating to both social and economic concerns and the implementation of policy at the most appropriate level. This came in the form of the EU's Fifth Environmental Action Plan (5EAP). Designed as a strategic document, the plan ran from 1992 to 2000 and sought to integrate environmental issues into mainstream EU policy-making on economic and social issues. It also highlighted the role of 'subsidiarity', where policies are implemented locally according to specific circumstances. The five key sectors for attention related to agriculture, energy, industry, transport and tourism. In 2001, the Sixth EAP began, running until 2010, which has emphases on climate change, nature and biodiversity, environment and health, and natural resources and waste. A key aspect of the 6EAP is the enforcement of existing environmental legislation.

Outside of these EAPs, the day-to-day process of EU legislation is focused around Directives, which set targets for environmental protection, which states are obliged to enforce in the best way that they see fit. A good example of an EU Directive that has had a major impact on policy-making at national, regional and local level is again the EU Landfill Directive (1999/31/EC) briefly discussed earlier in this chapter. This Directive asserts that by a given date (2015 for the UK), biodegradable waste being sent to landfill for disposal must be at 35 per cent of 1990 levels. Nationally, the government has produced a Strategy for implementing the Directive (DETR 1999a) and this will be reflected in regional waste strategies and, most notably, be reflected by the actions of local authorities in how they encourage the movement of this type of waste from landfill to other means of recycling or disposal (such as composting). Such specific Directives are often written within the wider context of Framework Directives, related to the strategic direction of policy in a specific sector, such as waste or water. Although very few exist in relation to the environment, EU Regulations are binding in their entirety and must be enforced by all EU nations.

EU policy reflects a weak interpretation of sustainable development (Connelly and Smith 2003) and has taken an overtly technocentric approach towards delivering sustainability. The reliance on economic instruments to effect change appears to reduce the role of individual citizens in change for sustainable development. However, the very nature of the EU as an institution means that its scope to promote effective policies for public engagement in sustainable development is limited. Nonetheless, the EU does recognise the important role of individuals in the sustainability debate. 'Agenda 21 has consistently highlighted the essential role of citizen involvement in all levels of environmental decision-making in guaranteeing the success of sustainable development' (CEC 1997, 112). This emphasis on the role of individual citizens is likely to be increased in the future given the current political climate within Europe. Recent rejections of the proposed EU Constitution by a number of member states has highlighted the importance of citizen engagement to the European Commission. This goes to the heart, in many cases, of questions relating to the role of the EU in policy-making and why nation states need to be part of a community of nations at all. From the environmental and sustainability perspective, the EU is important in both producing and enforcing environmental legislation, but at an international level it can also provide the framework for representing a series of national interests when negotiating with major economic blocks such as the United States. This has become evident recently with the pressure the EU has exerted on the United States over signing climate change agreements. Accordingly, although the EU is a remote institution for many individuals, it provides the framework for many of our most well-known environmental policies and therefore helps us to rationalise the nature of national policy-making.

National Policy for Sustainability: The United Kingdom

National policy, either in the form of a single strategic plan or a series of 'proofing' measures, is significant in terms of the overall direction offered by governments on how sustainable development should be interpreted. In the UK, this can be represented as a process, moving from initial reaction to the WCED report, to the current strategy published in 2005. This process is marked not merely by shifts in emphasis, but political changes that saw a change in government from Conservative to Labour in May 1997. This change had significant implications for sustainability policy, in particular how the government interpreted the very notion of sustainable development. Indeed, such changes reflected the shift in government policy from a conventional 'top-down' approach to a greater reliance on participatory mechanisms for sustainable development. As will be recounted below, the emergence of national sustainability policy is partly the emergence of a new politics for public engagement in sustainability, reflected in the Labour government's wider social inclusion agenda. Such an agenda has undoubtedly offered greater opportunities for an increased engagement of citizens in debates and action for sustainability.

In 1988, the Department of the Environment published a response to Brundtland's (WCED 1987) report (DoE 1988). What characterises the document is its positive nature: 'In essence, the Report is positive and optimistic, concluding on the note that there is hope if action is put in hand without delay' (DoE 1988, 11). Reflecting what Munton (1997 153) has termed a "market measures over regulation" approach, the response to Brundtland emphasised '... a new era of economic growth and the integration of environment and development' (DoE 1988, 11). The key word to note here is 'growth'. The Conservative government's response was a reflection of their wider political agenda, which emphasised the ability of market forces to create the conditions for greater efficiencies in the economy, leading to environmental protection. If we reflect on the principles of sustainable development that were examined in Chapter 2, the response of the UK government can be situated within the context of 'weak' and 'strong' approaches to sustainable development as this quotation illustrates: 'The rejection of zero growth is welcome ... The Report rejects the 'preserve everything' approach to conservation, recognising that growth and development involve changes...In the view of the UK, this emphasis on selectivity ... is the most realistic course' (DoE 1988, 11–12). This statement clearly rejects the strongest form of sustainability, implying that there are trade-offs to be considered in the determination of policies. Such a proclamation was indicative of the Conservative government's approach to environmental protection in the 1980s and 1990s (DoE 1989, 1990). Yet what is most striking about the response to Brundtland was the caution with which the UK government viewed any coordinated international programme to promote sustainable development, arguing that existing international structures were sufficient to review progress.

The immediate impact of Brundtland in Britain was the publication of DoE's White Paper on the environment, *This Common Inheritance* (DoE 1990). The Conservative government used this environmental strategy to discuss sustainability, inevitably framed in ecological rather than social or economic terms. This initial interpretation of sustainable development, from an environmental standpoint, highlighted three major lines of thinking in government policy for sustainability (Munton 1997):

1) the importance of shared responsibility for the environment, highlighting the role of citizenship and local initiatives;
2) the significance of a cross-cutting Whitehall agenda, involving a range of government departments;
3) the relevance of market forces to promoting environmental protection and economic efficiency. In essence, environmental protection was to be viewed as a further means by which to promote economic growth through efficiency, but equally must not restrict growth.

To this end, the publication in 1994 of the UK's first sustainable development strategy (DoE 1994) was deemed to be 'a disappointment' (Munton and Collins 1998, 348). The cause of this negative evaluation surrounded the lack of actual *strategy* in the document. As Munton and Collins (1998) have argued, the document re-produced many of the core principles in the environment white paper and emphasised current rather than future policy directions. Broadly, Munton and Collins (1998, 348) argued that 'The primary tone is one of caution; protecting the environment must not stand in the way of national wealth creation or economic competitiveness. The market is seen as the most effective mechanism for achieving more sustainable behaviour among both producers and consumers'.

The principles for the Strategy were outlined as:

* the precautionary principle;
* policy based on scientific evidence;
* shared responsibility between government, business, local authorities, NGOs and individuals;
* environment policy integration in Whitehall;
* the significance of the market over regulatory measures;
* a focus on growth and wealth creation.

It is worth pausing to consider the significance of the political philosophy underlying this set of principles. Most importantly, the principles reflect a political ideology that emphasises the importance of innovation and freedom to derive solutions to social and environmental problems, arguing that 'big' government and regulation acts as an inhibitor to economic and social development. In the three years that the Strategy was effectively operational under the Conservative administration, the dominant changes undertaken reflected this agenda and were inevitably focused around environmental dimensions:

* the integration of sustainable development into the lexicon of a large number of government departments and reflected in their own environmental strategies (Munton and Collins 1998);

- the formation of a single Environment Agency, as a result of the wide-ranging Environmental Protection Act in 1995 (Connelly and Smith 2003);
- re-orientation of the planning system for sustainable development goals (Munton and Collins 1998);
- economic decision-making with environmental considerations, particularly focused around valuation of the environment and the role of regulation versus market forces (Munton 1997);
- institutional changes (Munton and Collins 1998), leading to the formation of two government advisory bodies: the UK Panel on Sustainable Development and the UK Round Table on Sustainable Development. In addition to these bodies, a coordinating committee on the environment was established at ministerial level, alongside a set of 'Green Ministers' in government departments.

These moves forward were significant and had major advantages, yet the overall assessment of the 1994 Strategy has to be one of a 'missed opportunity'. The strict interpretations placed on sustainable development inevitably meant that wider issues of social change and empowerment were all but ignored in the document. Overall, the Strategy was, in a political sense, conservative – a cautious and institutionalised response to an agenda that was arguably radical and challenging.

New Labour: New Strategy?

In reflecting on the political climate of the mid-1990s in the UK, it is astonishing that such progress was made in relation to environmental policy given the wafer-thin majority that John Major's Conservative government was working with in the House of Commons. As leader of Labour's opposition, Tony Blair had already begun to set out a new agenda for sustainable development before his election in 1997, although it cannot be said that sustainability or the environment were major planks of Labour's successful 1997 election campaign.

The election of Tony Blair's Labour government in May 1997 was greeted by O'Riordan (1997a) with enthusiasm. The emphasis in the election campaign on a social justice and community evoked a feeling, reflected by O'Riordan (1997a, 12), that Labour would provide a radical alternative to the cautious approach adopted by the Conservatives towards new modes of governance: '... the nation is ready for a radical rethink of governance and of political principles, which could make the transition to sustainability a reality'. Yet the early months of government and the reality of being in power subsequently led O'Riordan (1997b) to become suspicious of Labour's intentions, labelling them 'green-ish'. This related partly to the political realities of government, which meant that for O'Riordan (1997b, 4–5): 'Promises are the froth of politics. The body lies in the programmes ... Reader, please do not hold your breath. This is going to be a long haul'.

Such suspicion arose partly from the evident recognition by ministers that Labour had been elected on a platform of delivering good economic governance and an understanding that the stewardship of the economy was a high-profile issue in the electorates mind. To this end, more radical policies such as re-nationalisation of the railway network had been quietly dropped. An easier pill to swallow and one which played equally well to both traditional and new Labour voters was the emphasis on the perceived lack of social concern expressed by the previous administration. Accordingly, in 1998 Labour published the consultation document for a new sustainable development strategy. This was focused around what the Deputy Prime Minister termed 'A historic programme to build a modern Britain, a strong economy and a healthy environment to pass on to our children' (DETR 1998, 1).

The shift in emphasis from the previous government's policies was clearly evident. Key words and phrases, now recognised as central to the discourse of New Labour, littered the document – social justice, quality of life, future generations. The focus on people was significant and was reflected in the initial priorities set out in the consultation document, in this order:

- sustainable goods and services;
- development of sustainable communities;
- management and protection of environmental resources;
- sending the right signals as a government;
- the importance of international action.

Opportunities for Change (DETR 1998) therefore represented a step-change in government attitudes towards sustainable development, highlighting the importance of linking social dimensions of sustainability (particularly social justice and social inclusion) to the management of and access to environmental resources.

Publication of the 1999 strategy, *A Better Quality of Life* (DETR 1999b) therefore represented many of these sentiments. Unlike the principles outlined in the 1994 Strategy, this document focused on key issues for concern and action. These focused around the following four aims:

- social progress, which recognises the needs of everyone;
- effective protection of the environment;
- prudent use of natural resources ;
- maintenance of high and stable levels of economic growth and employment.

These aims presented what many had desired in a truly sustainable strategy in terms of the integration of social, environmental and economic goals. The holistic and all-encompassing nature of these aims was therefore significant. Also important was the order in which they were presented. The priority afforded to social progress recognised the reforming aspirations of New Labour, with a mandate to tackle social exclusion and inequality in British society. The next two aims – effective protection of the environment and prudent use of natural resources – were in many ways noncommittal, but within the context of Pearce's (1993) sustainability spectrum (Figure 2.6), would certainly be representative of a weak sustainability position, given the lack of reference to critical levels of natural capital. Finally, the use of the word 'growth' in the fourth aim also points to a wider political context. What the Deputy Prime Minister John Prescott termed 'wise growth' (rather than 'sustainable development') made reference to the central need for growth, reinforcing the impression that the Strategy was a weak interpretation of sustainable development.

This weak approach to sustainability that appeared to be evident in the Strategy was made more explicit by the principles under which the Strategy was to operate:

- the *precautionary principle*: objective assessment of risks that have both negative environmental *and* social impacts;
- the definitions of *capital* in the Strategy:
  - ○ social: skills, knowledge, health, self-esteem, social networks and communities;
  - ○ environmental: implicit need for trade-offs.

Although the precautionary principle had been used in numerous national environmental strategies in the past, the balancing of social and environmental impacts implied the same ordering of these two sectors, which became evident when the notion of capital was addressed. The need to acknowledge trade-offs between environmental capital and social and economic capital reinforced a view that the Strategy was focused more acutely at social, rather than environmental concerns.

Perhaps the most significant departure from the 1994 Strategy was the central role afforded to the measurement and evaluation of progress towards sustainable development. This came in two forms: the development of *sustainability indicators* and a range of *institutional changes*. A great deal of space was given in the Strategy the explanation and justification for a set of *sustainable development indicators*. This focus on indicators would latterly become a sticking point for the Strategy, but in a burst of enthusiasm from a new government, a series of indicators were developed to measure progress towards sustainability. These were organised into five national indicator 'families':

- economic growth;
- environmental protection;
- social progress;
- international progress;
- government progress ('setting our house in order').

The fourth family reflected the new government's determination to place development issues higher on the political agenda. In total, 147 national indicators were developed as members of these five families. However, to chart overall progress, 15 headline indicators were instituted (Figure 3.3). These headline measures were designed to present a snapshot perspective on progress, rather than be representative of overall movement (DETR 1999c). Yet the apparent

Ref no.	Headline indicators
H1	Total output of the economy (GDP and GDP per head)
H2	Total and social investment as a percentage of GDP
H3	Proportion of people of working age who are in work
H4	Indicators of success in tackling poverty and social exclusion (children in low income house-holds, adults without qualifications and workless households, elderly in fuel poverty)
H5	Qualifications at age 19
H6	Expected years of healthy life
H7	Homes judged unfit to live in
H8	Level of crime
H9	Emissions of greenhouse gases
H10	Days when air pollution is moderate or higher
H11	Road trafic
H12	Rivers of good or fair quality
H13	Populations of wild birds
H14	New homes built on previously developed land
H15	Waste arisings and management

Figure 3.3   Headline indicators of sustainable development between 1999 and 2004 (DETR, 1999c). © Crown

Source: DETR (2004a).

specific nature of some of these presented numerous communication problems. For example, if the population of wild birds was declining, did this represent an overall degrading of biodiversity? As we shall see below, the inadequacy of such indicators would lead to a radical overhaul in the measurement rubrics used in sustainability evaluation.

A second measurement and evaluation system outlined in the Strategy was the need for institutional change within government and the accountability measures used to ensure progress. Two Whitehall initiatives are of significance here:

• Green Ministers and the ENV(G) aimed to:
   ○ consider the impact on sustainable development of government policies;

- ○ improve the performance of departments in contributing to sustainable development;
- ○ report as necessary to the Committee on the Environment, Transport and the Regions;
- SD Task Force:
  - ○ a mechanism for interdepartmental discussion of areas where the headline sustainable development indicators are not moving in the right direction.

The first of these initiatives was established under the Conservative administration, whilst the second was a new initiative providing a strategic overview of progress and was designed to act as a checking mechanism. More broadly, this represented a shift towards what has become known as the 'proofing' of policy for sustainable development in government departments. To this end, these programmes represented the growing ability of sustainable development to permeate government departments both horizontally and vertically. Nonetheless, the boldest move in terms of evaluating government progress was the establishment of the *Sustainable Development Commission*. Previous attempts by the Conservative government to formulate panels of expertise to form the basis for policy discussion had been centred on two forums: the *UK Round Table on Sustainable Development* and the *UK Panel on Sustainable Development*. New Labour considered that these two institutions had effectively run their course and lacked political bite. In the newly formed Sustainable Development Commission (SDC), Labour were seeking what the Strategy termed a 'critical friend'. Comprised of commissioners with a range of expertise and chaired by the well-known environmentalist Jonathan Porritt, the SDC held the following remit from government:

- review how far sustainable development is being achieved in the UK in all relevant fields and identify any relevant processes or policies which may be undermining this;
- identify important unsustainable trends which will not be reversed on the basis of current or planned action and recommend action to reverse the trends;
- deepen understanding of the concept of sustainable development, increase awareness of the issues it raises and build agreement on them;
- encourage and stimulate good practice.

A Real Opportunity for Change?

The 1994 Strategy was representative of a re-packaging of existing environmental policies, with a focus on economic growth and creating efficiencies in the economic system to protect the environment. Aside from the major institutional changes outlined above, the narrative of sustainable development and the discourses evident within the new 1999 Strategy were a major shift towards a socially-inspired agenda. Although economic growth was still the 'political reality', this was embedded in discourses that highlighted the need to balance economic progress with social imperatives and environmental considerations. From the perspective of this text, the increasing importance afforded to social dimensions was significant, marking a change in policy emphasis towards a model that recognised the link between social and environmental capital. This link implied that increasing levels of social capital, sustainable community development and quality of life would lead to greater levels of environmental activism. Such a hypothesis was drawn from what we will examine later in this chapter in the form of the 'new communitarianism'. However, suffice it to state that the new Strategy placed emphasis on both rights (the right to environmental quality) and responsibilities (the need to make choices and change behaviours as consumer to protect the environment for ourselves and future generations). Within the context of this text, the 1999 Strategy therefore represented an acknowledgement that individual social actors could have a central role in promoting sustainable development.

Taking It On: Current National Sustainability Policy

Almost immediately after its publication, the 1999 Strategy was receiving a range of critiques, partly aimed at its conceptual content and partly at the progress being made inside Whitehall. Porritt and Levett (1999, 5) led the critique in conceptual terms, arguing that the Strategy was timid, representing a sop to sceptics of sustainable development 'Let's make them all feel safe with the terrain instead of them seeing it as alien territory populated by tree huggers and little green men and women from another planet'. To this end, Porritt and Levett argued that environmental concerns were neglected over social imperatives. Indeed, whilst economic and environmental objectives had been integrated at what he termed 'level 1' (conceptually), there was little evidence of integration at 'level 2' given that each set of policies were 'pulling' in alternative directions. As concerning were the problems regarding progress within central government. Government was not seen to be "walking the talk" (Porritt and Levett 1999, 5), with the observation that although the Strategy contained numerous indicators of sustainable development, there was a notable lack of specific targets. These claims were also highlighted by both Ainsworth (2004) and Wilson (2000) who argued that sustainable development strategy within Whitehall was represented by a lack of resources on the one hand and a proliferation and duplication of policies on the other.

This growing unease culminated in a review of the 1999 Strategy at its five year milestone with both a report by the Department of the Environment, Food and Rural Affairs (DEFRA, formerly DETR) providing an assessment of progress in terms of the indicators developed in 1999 (DEFRA 2004a). At the same time, however, the Sustainable Development Commission also published its evaluation of these indicators (SDC 2004). The government's view was that progress was positive, but areas of concern remained (DEFRA 2004a). The very title of the Sustainable Development Commission's report *Shows Promise, but Must Try Harder* indicated the approach taken (SDC 2004, 9). In a frank review of progress, the Commission stated that:

... there is a significant gap between the Government's assessment of progress and our own ... Our own assessment is that neither the UK Government or its devolved administrations nor our society have as yet fully assimilated how far the goals of sustainable development represent a radical critique of present policies and achievements ... much more needs to be done in engaging society as a whole in facing up to the challenges of sustainability

The report presented an alterative view of sustainability, predicated on the basis of the 'wellbeing of society itself and the planetary resources and environment that sustains us all' (SDC 2004, 9). The centrality of individuals to progress was clear in the report, which argued that too little effort and progress had been made in raising the awareness of citizens to both the major problems posed by environmental degradation and the challenges of shifting consumer patterns to create a more sustainable society. In its 38-page report, the Commission outlined 20 challenges for the government, beginning with the scripting of a new strategy for sustainable development to include a new set of indicators, objectives, principles and institutional changes. Economically, the Commission called for adjustments to the tax system to account for environmental costs of economic activity, as well as the promotion of sustainable production and consumption through greater use of public procurement. Environmentally, key challenges were seen as reducing waste and tackling climate change. Socially, education for sustainable development was placed as a priority.

However, the greatest amount of criticism from the SDC was reserved for the type and use of indicators utilised by the government. The SDC (2004a) highlighted the difficulties endemic in using basic measures of economic performance (such as Gross Domestic Product or GDP). As they reported, since 1970 GDP has risen along a generally linear trajectory, with the exception of flatter periods during times of recession. Yet life satisfaction as recorded along the same timescale has not increased at all. Indeed, economic measures such as GDP also fail to take into account growing levels of consumer debt and the hardship that this can bring. Not only were the use of specific indicators criticised by the SDC, but the measures selected to reflect specific areas of performance were also examined. Although car ownership has increased year on year, presenting major environmental challenges, the growth in low cost air travel has far more significant consequences for the global environment. This raises the very question of what key priorities society wishes to establish for itself. The SDC noted that the 15 headline indicators represented measures of quality of life that focused on economic growth and performance. A wider set of indicators that attempted to capture a greater range of social experiences and 'needs' was necessary.

Responding to these challenges, the government set in train the consultation for a new Sustainable Development Strategy (DEFRA 2004b, 6), using the following definition of sustainable development: '... to enable all people throughout the World to satisfy their basic needs and enjoy a better quality of life without compromising the quality of life of future generations'. What was significant about this consultation document was the increasing prominence afforded to the 'big' challenges (DEFRA 2004b, 30). These focused around climate change, energy, sustainable consumption and production, environment and social justice, sustainable communities and changing behaviour. This final category demonstrated the growing recognition amongst policy-makers that sustained and significant change was only likely to occur through a commitment by the majority of individuals to a shift in daily practices and lifestyle changes. The government posed the question 'How is the UK likely to be most successful in achieving the behaviour changes that will be needed if we are to move towards long-term sustainability and what would be the right balance of measures by government and others?' Publication in 2005 of *Securing the Future* (DEFRA 2005) presented what can only be termed a further step change. The document had four major elements of note:

- a stronger international focus;
- five guiding principles;
- four key priorities;
- a new indicator set.

The wider *international focus* provided by the document was a further example of the Labour administration's attempts to view the UK economic system within the context of a globalised trade network and reflected emerging discourses of free trade and debt relief that would culminate in commitments to cancel large amounts of developing world debt in the July of 2005. The *guiding principles* presented five core policy emphases (Figure 3.4). The hierarchical nature of the diagram is extremely important to note. Conceptually at least, the lower priority afforded to environmental protection in the 1999 strategy (Porritt and Levett 1999) was replaced by two key 'end points' for sustainability policy: 'Living within Environmental Limits' and 'Ensuring a Strong, Healthy and Just Society'. The economic objectives were downgraded to the role of feeding into these two core objectives. Accordingly, the importance of placing environmental limits and a sustainable society at the centre of sustainability policy appeared to cement the notion that the economy had to operate within specific environmental constraints, reducing the potential for trading critical natural capital for other capital.

This emphasis on environmental limits and sustainable communities was also illustrated by the four key priority areas identified:

- sustainable consumption and production;
- climate change;
- natural resource protection;
- sustainable communities.

In the Strategy document, chapters were used to develop these priorities further, but what preceded these sections was a contextual chapter seen as the basis for these four priority areas – *Helping People Make Better Choices*. This chapter and its contents were regarded as central to achieving sustainable development. As the Prime Minister, Tony Blair, indicated in the foreword to the Strategy (DEFRA 2005, 3): 'Each of us needs to make the right choices to secure a future that is fairer, where we can all live within our environmental limits. That means sustainable development'. Chapter 2 begins with the statement that (DEFRA

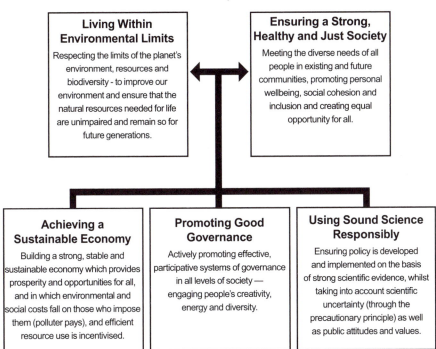

Figure 3.4   Guiding principles from *Securing the Future* (adapted from DEFRA 2005). © Crown

Source: DEFRA (2005).

2005, 25): 'Behaviour changes will be needed to deliver sustainable development'. The chapter outlines five key approaches to delivering behaviour change. First, the document outlines a framework for conceptualising change (Figure 3.5). Within this notion of behaviour change, government sees four ways to provide a catalyst for positive behavioural adjustment, through the use of social engagement, leading by positive example, economic and social encouragement and the enabling of activity by the removal of barriers and provision of facilities. It is interesting to note that this framework is used throughout the document as the means by which to effect specific policy outcomes, but that it is grounded in the basic behaviour change mould. Subsequently, the chapter outlined the means by which such measures will be implemented. As a means of creating more engaged communities, *Community Action 2020: Together We Can* was outlined. Spearheaded by the Home Office, this initiative was reflective of wider government initiatives to reverse the downturn in community participation. It sought to:

- build safer communities;
- reduce re-offending;
- support young people's development;

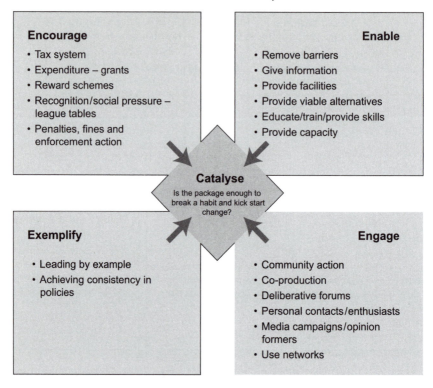

Figure 3.5   **A framework for behaviour change presented in** *Securing the Future*
(adapted from DEFRA 2005). © Crown

Source: DEFRA (2005).

- increase community cohesion;
- strengthen democracy;
- revitalise neighbourhoods;
- make the best use of schools and improve public health.

Although an environmental agenda is not evident within this framework, the link between greater social and community capital and environmental sustainability through increased ownership of resources and initiatives lies at the heart of this programme.

In terms of communication of sustainability messages, the chapter makes clear that previous approaches to behaviour change have resulted in a notable rise in awareness, but a negligible change in behaviour (see Chapter 4). Raising awareness is still given prominence, but there is an acknowledgement that behaviour change is only likely to result if messages are tailored to specific groups in society and related to existing hopes and lifestyle aspirations. Two other mechanisms for behaviour change are also outlined: economic instruments (in the form of taxes or incentives) as well as the use of education in schools for sustainability. The focus in this text is on the use of engagement and behaviour change strategies to effect action without the use of economic instruments and is focused on the adult population.

A final and significant element of the 2005 Strategy is the emphasis placed on implementation. Delivery mechanisms are outlined and 20 new 'framework' indicators developed on the basis of the evaluation made by the SDC (DEFRA, 2006), with a reduction to 48 of the total number of national indicators (Figure 3.6). These indicators are more representative of the specific priority areas defined by the government and reflect the need highlighted by the SDC to attempt to measure more meaningful aspects of sustainable development, such as well-being and quality of life, rather than simply economic growth. The SDC itself will also have a more enhanced role, with increased resources and will act as the body responsible for assessing progress of the Strategy.

National Sustainability Policy: A Summary and Assessment

Sustainable development policy has travelled a long way since the initial responses to the Brundtland Report (DoE 1988). The major shifts in policy emphasis have been threefold:

- concepts: a shift from a weak approach to sustainable development, emphasising an ecological modernisation perspective, to a strong*er* approach that now recognises, in principle, environmental limits and the role of critical natural capital. To this end, we might state that there has been a shift along the technocentric-ecocentric spectrum towards the latter. However, the dominant discourse of policy is still economic growth;
- priorities: a move away from a purely economic imperative towards a focus on environmental limits and community development. The current situation has been through a number of iterations, with an initial focus on economic growth as a mode by which to effect environmental protection, changing to a perspective dominated by a socially-driven agenda and finally emerging with a contemporary emphasis on key environmental and social issues;
- practices and implementation: a radical change from a technocratic perspective emphasising the role of science and market economics to effect change to a socially-driven agenda that places individual behaviour change as a means to implement policy.

These changes have been undertaken within the wider context of the growing proliferation of sustainable development into an ever-expanding range of government departments and initiatives. In attempting to evaluate the overall Strategy, the SDC (2005) provided an initial reaction in late 2005. There was significant praise for the tone of the new Strategy, not least the emphasis on limits and the new indicator set. Indeed, the focus on 'well-being' as a concept was particularly welcomed as an alternative to economic growth as a measure of progress. There was also praise from the SDC for the way in which the government had attempted to deal with the vexed issue of behaviour change. Yet the SDC were

Environment and Society

Indicator

Greenhouse gas emissions	
Resource use	
Waste	
Bird populations:	*Farmland*
	Woodland
	Coastal
Fish stocks	
Ecological impacts of air pollution:	*Acidity*
	Nitrogen
River quality:	*Biological*
	Chemical
Economic output	
Active community participation	
Crime:	*Vehicles and burglary*
	Robbery
Employment	
Workless households	
Childhood poverty	
Education	
Health inequality:	*Infant mortality*
	Life expectancy
Mobility:	*Walking/cycling*
	Public transport
Social justice	
Environmental equality	
Wellbeing	

Figure 3.6   The 20 framework indicators for sustainable development.
© Crown

Source: DEFRA (2006).

still unclear how the 'big' issues of climate change and debt relief would actually be tackled effectively. To this end, questions were raised over the implementation of the Strategy and its effectiveness at the regional level. Having examined the national perspective on sustainable development, it is now time to turn to the regional and local scales.

The Regional and Local Dimensions

National policy provides a framework for action at more refined geographical scales, yet Agenda 21 (UNCED 1992, ch. 28) highlighted the role for local authorities (at varying scales) as being the most appropriate level to implement sustainable development. Chapter 28 of *Agenda 21* argued that local authorities needed to provide:

- institutional consultation for local governance;
- public consultation in local areas;
- information transfer between local areas;
- inclusivity.

Democratically, local authorities are more accountable to local people, but they also provide strategies that will be of greater relevance to local stakeholders. To this end, the potential for encouraging participatory mechanisms for tackling sustainable development should become greater at this level.

Regionalising Sustainability

Since the 2004 Planning and Compulsory Purchase Act, the way in which local government operates has begun to alter significantly. Previously in England and Wales, a two-tier system of planning in local government operated (Gilg 1996) with decisions on local issues being split between district and county authorities. In a number of locations, 'unitary status' has been granted to local authorities, condensing all of the functions into one super-authority. The county has conventionally held responsibility for overall strategic policy, including planning. Under reforms brought in during 2005, the role of the region in the planning process has been increased in status, with counties occupying a more conspicuous and questionable space. Gradual changes towards a 'regionalisation' of local government have been undertaken during Labour's term in office and Figure 3.7 presents the government's representation of how this new system works. The regional level is comprised of three distinct bodies:

- Government Offices for the Regions (GORs): whose role is to represent government departments in the regions and to ensure that sustainable development policies in these departments are being undertaken regionally, as well as reflecting regional concerns to central government;

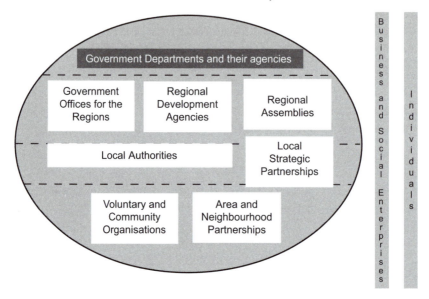

Figure 3.7   National, regional and local delivery structures for sustainable development in the UK (adapted from DEFRA 2005). © Crown

Source: DEFRA (2005).

- Regional Development Agencies (RDAs): who have a remit for promoting economic development in their respective region and for compiling a Regional Economic Strategy. They have a statutory obligation to contribute to sustainable development in the region;
- Regional Assemblies (RAs): who scrutinise the work of RDAs, as well as preparing Regional Spatial Strategies (RSSs) and Regional Sustainable Development Frameworks (RSDFs).

As the main leading role and voice for sustainability in the regions, Regional Assemblies (RAs) play a key role in negotiating RSDFs between key stakeholders in the regions. The RAs also have a broad strategic role in coordinating the myriad of regional strategies in other sectors of their remit, such as transport, health, tourism, etc. A good example of this coordinating and overall strategic role is the South West of England's Regional Assembly Integrated Regional Strategy (IRS) for the period 2004 to 2006 (SWRA 2004). Within this document, more than 18 strategies feed into the Regional Sustainable Development Framework (RSDF), creating a major task for those seeking to provide a voice to diverse economic, social and environmental interests. Indeed, the South West's geographical scope (from the Isles of Scilly to Gloucestershire) and its varying socio-economic base make the task of managing the RSDF a significant problem for regional policy-makers.

    Regional Assemblies also provide the link for regional policy to the local level. In formulating policy, they take account of the local strategies that are being developed, such as Community Strategies. Under new planning regulations, the hierarchical nature of the policy process can be represented as shown in Figure 3.8. Regional policy, in the form of the IRS, RSDF, RSS and other regionally-based strategies form the overall framework for making policy locally. Within local authorities, the main driver for implementing these policies is the Local Development Framework (LDF) informed by the Community Strategy (see below). Local Development Frameworks set the basis for localised development to reflect the regional strategy. However, it must be recognised that the process is cyclical, with Community Strategies, for example, feeding into regional planning processes.

Figure 3.8   Sustainable development: the planning policy framework

Although the move to regionalisation has been progressing at a steady pace, there are a number of questions that arise from the increasing regionalisation of sustainability policy. Not least, there is seen to be a significant democratic deficit in the way that RAs, as the bodies that compile a plethora of strategies which have major implications for communities. RAs are appointed from existing elected representatives from local authorities, but also from a range of stakeholders selected to represent specific interests. The relative lack of democratic accountability at this level raises concerns that were less evident at the county level given the ability to elect county councillors[1] (although it should be noted that voters in the North West of England chose not have an elected regional assembly).

A further concern highlighted by the regionalisation of British local government is the ability of 'the region' to develop sound and sustainable strategies and policies for the region as a whole. The lack of spatial cohesion and specificity raises questions over the ability of one body to formulate policy for a diverse geographical

---

1   As this book was going to press, the Prime Minister announced that Regional Assemblies were to be disbanded, with their role being split between local authorise and the Regional Development Agencies. This 'split' in responsibilities was not finalised, although the enhanced role for RDAs implies that the role of the region will still remain significant.

region. In the South West, for example, the area covered by the regional assembly stretches from the Isles of Scilly to Gloucestershire. Both this and the previous concern raise the question of whether it is possible for the region to act as a viable conduit for sustainable development policy. Indeed, given the recent nature of such developments, it is too early to give a rigorous account of the potential for regional administration to effectively create meaningful and distinctive sustainable development priorities. Indeed, most experience of implementing sustainable development has undoubtedly been at the local authority level, to where we now turn our attention.

Localising Sustainability: Local Agenda 21, Community Strategies and the 'New Localism'

The agenda to promote sustainable development at the local community level began immediately after the Rio Earth Summit. In Chapter 28 of *Agenda 21*, the nature of a Local Agenda 21 was clearly outlined:

> Each local authority should enter into a dialogue with its citizens, local organizations and private enterprises and adopt 'a local agenda 21'. Through consultation and consensus building, local authorities would learn from citizens and from local, civic, community, business and industrial organisations and acquire the information needed for formulating the best strategies. (UNCED 1992, Chapter 28)

More specifically, Selman and Parker (1997) argued that Local Agenda 21 presented an alternative to conventional forms of community development for two major reasons:

- a shift from 'official' policies, management systems and targets to significant cultural changes in local administration and agencies;
- it necessitated a shift in public attitudes and behaviour.

In essence, Local Agenda 21 was about a significant cultural shift in attitudes towards the very way in which local areas functioned and lived. This needed to give voice to 'others' within the community and generate a commitment by the majority to alter existing patterns of activity. The Local Government Management Board (LGMB 1994) identified six areas where practical actions were required:

- improving local authority environmental performance;
- integration of sustainable development aims into local policies and strategies;
- raise awareness and education about sustainability;
- consultation of the public;
- establish new partnerships across all sectors of the community;
- measuring and mentoring progress towards sustainability.

Embedded in the *Fifth Environmental Action Plan* (EC 1992), the promotion of local sustainable development was regarded as the most effective means by which to make major changes that would have both global and local impacts. As we shall see in this section, the emphasis during the first eight years or so was (somewhat inevitably given the tone of the Rio Earth Summit) a focus on local *environmental* sustainability. In many cases, local sustainability initiatives represented local environmental initiatives that had been re-born with an alternative title. Indeed, this may have been part of the problem in engaging more than a minority of individuals to participate in such programmes. One of the most common mantras for local sustainable development became 'Think Globally, Act Locally' (Steel 1996). As many of those reading this text will know, this phrase has now been replaced by 'Think Locally, Act Locally', representing the moving discourse of local sustainable development from a series of concerns generated by a global environmental 'crisis' to a myriad of challenges from waste management to tourism. Yet as we shall also see, in tackling the participatory agenda for sustainable development, it is at this local level that policy formulation has been most innovative, attempting to blend conventional and traditional notions of environmental concern with new ideas of what a sustainable community could look like and the behaviours necessary to achieve such goals. Accordingly, we will firstly examine the institutions for local sustainability that have developed, both at the international and local levels, moving on to assess the local structures for sustainability and the promotion of Local Agenda 21 programmes and finally examining how local sustainability initiatives have succeeded in promoting participation in sustainable development and the extent to which such participation is 'sustained'.

Institutions for Local Sustainable Development

Two dominant forces can be identified in determining the relative high profile of Local Agenda 21 policies in the UK during the 1990s. Nationally, the Local Government Management Board (LGMB) has been a consistent advocate of such local initiatives and in a statement to the UNCED before the main conference at Rio in 1992, it argued that a '… local democratic mandate enable local authorities to inform, mobilize and speak on behalf of their authorities' (LGMB 1994, 31). The LGMB provided the overall strategic vision and drive to see a local agenda 21 process evolve in the UK. However, much of the intellectual and practical drive emerged from the International Council for Local Environmental Initiatives (ICLEI). ICLEI has become known internationally for providing both a means of collecting examples of good and best practice on environmental practices, as well as fulfilling a lobbying body for the role of local environmental programmes. ICLEI has encouraged the use of 'Strategic Services Planning Frameworks, which can be used to provide both an inclusive approach to plan-making and delivery, as well as an evaluative framework.

Within these two frameworks, the business of achieving the UNCED's aspiration that 'most' local authorities should have instituted a Local Agenda 21 process by 1996 was mainly left to these bodies to fulfil. The 1994 Sustainable

Development Strategy (DoE 1994) did not make specific reference to a date by which UK local authorities had to complete the process. To this end, local authority decisions on how and at what speed to proceed were driven mainly from the LGMB and more indirectly from ICLEI. In 1997, after the election of a new Labour government and the formation of the Department for the Environment, Transport and the Regions (DETR), a more formalised process of LA21 became instituted. This was driven by the Prime Minister's declaration in 1997 that by the year 2000 all local authorities must have completed a consultative process and have published a Local Agenda 21 plan. In the following section, we will examine studies by Tuxworth (1996) and Morris (1999) to establish the extent to which progress was made. However, in 2000 the institutional framework for Local Agenda 21 began to change in a move towards what have become known as Community Strategies (Local Government Act 2000). In moves to amend the planning process and place greater emphasis on community planning within the general planning framework, central government argued that the often sidelined and narrow view of Local Agenda 21 should be broadened to include a range of community aspects. This emerged partly from the assertion that many Local Agenda 21s were little more than well developed *environmental* strategies (Tuxworth 1996). Community Strategies now provide many of the functions of LA21, but do so within a strategic local authority planning framework and have rationalised and institutionalised the consultation and development processes originally outlined by the LGMB (1994). Accordingly, the scope of LA21 has now been subsumed into the wider community development process and because of the relatively recent nature of this change, the remainder of this chapter focuses on an evaluation of Local Agenda 21 as a process for public participation and local change. One point to bear in mind when reading the following sections is that one of the major criticisms levelled at Local Agenda 21 was the environmental focus it contained. Community Strategies have attempted to addressed this apparent unbalance (Ross, 2000). However an over-arching and as yet unanswered question is whether the all-encompassing nature of Community Strategies now means that any tangible environmental message has been lost (Ross 2000). The introduction of the latest government campaign for community action (Community Action 2020) may address this issue as the current sustainable development strategy is worked through. Indeed, DEFRA (2004b) argued that Community Action 2020 was needed to address fundamental mis-understandings of sustainability at the local level.

The Reality of Local Agenda 21

During the 1990s, the Local Government Management Board (LGMB) commissioned a number of studies to examine the extent and nature of Local Agenda 21 plan production. Tuxworth's (1996) study revealed a series of trends in the development of LA21 which, given the significance of the date, are worthy of note. It is important to qualify these generalisation with the qualification that the response rate was only just over 50 per cent:

- by 1996, 78 per cent had made a commitment to produce an LA21;
- those not committed blamed either lack of time to work on LA21, or a corporate unwillingness to participate;
- responsibility for implementing Local Agenda 21 was most often given to existing staff to manage within their current role;
- LA21 was most often discussed and debated through existing local authority structures;
- the vast majority of local authority interests served by LA21 were environmental initiatives;
- although conventional means of engagement and consultation were evident innovative methods such as 'planning for real exercises' or 'visioning' exercises were rare.

These conclusions highlighted the difficulties of attempting to integrate LA21 into existing local authority structures and also the difficulties of communicating central message of sustainability, which related to both social and economic issues, as well environmental concerns. After the Blair declaration of 1997, a further survey undertaken by Morris (1999) attempted to gauge progress towards the 2000 deadline for plan production. With a response rate of 77 per cent, 81 per cent stated that they had produced or would have produced a strategy by 2000. Morris highlighted some key trends of note:

- a disappointing level of involvement in the LA21 process from women and ethnic minorities;
- the 'vision' produced for a locality often had a narrow consultative basis;
- the dominant areas of work for LA21's were still environmentally-focused, such as home energy conservation, biodiversity and land use planning.

These two studies highlighted a range of challenges inherent in the LA21 process, which relate to (1) institutional commitment and administration, (2) focus and (3) participation. In terms of institutional and administrative issues, Tuxworth's (1996) analysis provides the empirical evidence in relation to the integration of LA21 into existing local authority structures and the challenges faced by corporate inertia or hostility. However, from a more qualitative perspective Gibbs et al. (1996; 1998) have provided an in-depth study of how interpretations of sustainability can influence the direction of local sustainability practice. Using Pearce's (1993) sustainability spectrum as their guide Gibbs and his team identified four types of local authority responses to sustainable development:

- *very weak*: modification in existing structures; surface appearances and minor changes;
- *weak*: some process changes; less tangible problems dealt with;
- *strong*: system changes as a whole, examining the 'system' as one element;
- *very strong*: a 'cultural' change; external as well as internal elements of the system altered.

The main finding of the research was that the interpretations given above depended on where the officer charged with responsibility for LA21 resided. Indeed, there was a major division according to the aspirations of LA21 programmes, determined along an environment – economic dichotomy. For those authorities where the LA21 officer was based in the economic development directorate, sustainable development was viewed as a means by which to enhance economic growth and investment. What Gibbs et al. (1996) term 'surface changes' were indicative of paying 'lip-service' to the environment or using environmental initiatives to enhance economic performance. Conversely, those LA21's and their respective officers located in environment directorates presented a set of proposals that saw their local authority within the context of a global environment, with significant emphasis placed on the global commons. Examples of this line of thinking emphasised the role of reduced car use in contributing to the global fight to tackle climate change. Overall, therefore, there was a feeling that the local authority had to make a contribution to a 'global cause'. Accordingly, the institutional commitment and interpretation of sustainability had a significant impact on how policies and plans developed. Indeed, as Selman (1996) pointed out, the effect of institutional concerns did not impact merely on the local authority's perspective of sustainability, but also the practices used to formulate strategies. As the LGMB pointed out (LGMB 1994), partnership working between local authorities, built around consensus and consultation, was vital for the success of LA21s. However, citing work from the Department of the Environment, Selman (1996) highlighted the negative features that emerged in the 1990s during attempts to formulate partnerships, not least the sectoralism, mutual distrust and different languages used by alternative sectors. Sustainability therefore presented some very unique challenges, posing problematic questions relating to social and economic priorities, which alternative interests had to negotiate in the process of plan formation. As Selman (1996) pointed out, these tensions were partly the reason for the alternative models of implementation that were adopted by a range of local authorities. For example, some local authorities (such as Gloucestershire) handed the entire process over to a voluntary organisation. Other authorities (e.g. Leicester) were able to formulate very strong partnerships for action, whilst many (e.g. Exeter) were led solely by the local council.

The Focus of LA21

Beside these institutional and administrative issues, a major concern for researchers of LA21s in the 1990s was the overtly environmental focus they took (Agyman and Evans 1994; Freeman et al. 1996; Selman, 1996; Selman and Parker 1997). There are numerous examples of this environmental focus (see Tuxworth (1996) and Morris (1999) for quantitative examples). However, an examination of a specific LA21 can assist in illustrating this point. Exeter's LA21 was agreed and published in 1996 (Exeter City Council, 1996). Aesthetically, the document provides an overtly 'green' image, illustrated in Figure 3.9 by the logo, using the image of a globe (the major symbol of which is a plant) being supported by a

Figure 3.9   Local Agenda 21 logo from Exeter City Council's LA21 Strategy (adapted from Exeter City Council 1996). © Exeter City Council. Reproduced with permission

Source: Exeter City Council (1996).

hand. This visual representation of sustainable development is explored textually by two quotations at the beginning of the document:

> ... living and developing in a way which ensures that future generations can enjoy the same quality of life as we do today (Exeter City Council 1996, 2)

> ... we do not inherit the Earth from our parents, we borrow it from our children (Exeter City Council 1996, 2)

These definitions are reflected in the priorities that the Council saw for the city: '... clean air to breathe; a clean and wholesome food supply to eat; decent housing; and easy and safe access to green spaces and community facilities' (Exeter City Council 1996, 1). The Strategy is conspicuous for its lack of focus on economic issues; the dominant themes are the related issues of environmental protection and community development. Exeter's consultation process was based on a consultation with 100 interest groups, 2000 residents, an Agenda 21 exhibition and local publicity. The results of the consultation highlighted the priorities given in the quotations above. Nonetheless, the overtly 'green' focus of the literature produced by the Council appeared to lend an environmental focus to the process. This is reflected in the 16 indicators of sustainable development for the LA21, which contained nine environmental indicators and seven social indicators, but with no economic reference point.

An LA21 such as Exeter's was very common during the 1990s. No one can doubt the integrity or commitment of the LA21 officers involved in the production of such documents. However, their focus on an environmental agenda was significant because it gave the general impression that LA21s were extended and more ambitious environmental strategies – another way to encourage people to recycle and save energy – but little more than this in terms of the wider social and economic agendas.

LA21 and Participation

Perhaps the most fundamental issue for LA21 has been the significance it afforded to community and individual participation. Seen as central to the LA21 process and more widely to the cultural shift identified by Selman (1996), the emergence of participatory mechanisms for sustainability emerged at an interesting point within the political process in Britain. The emergence of an invigorated Labour party, firstly in opposition and then in government, was leading to a re-examination by politicians of the role of citizenship in society. Theoretically, the intellectual agenda was driven by a guru of Tony Blair's, Amatai Etzioni, who had highlighted the importance of moving towards what he termed the 'new communitarianism' (Etzioni 1993). This was a recognition within society of the role that both rights and responsibilities played in cementing community relations, enabling a 'majority' to compel a 'minority' to engage in socially desirable behaviours. This agenda has been taken forward by Tony Blair's government in the form of citizenship teaching within the national curriculum. The notion of citizenship within an environmental context utilises the basic tenets of communitarianism, highlighting the role that rights and responsibilities play (Waks 1994). Selman (1996) has argued that sustainable development necessitates the movement towards a new form of environmental citizenship, with access rights to a clean and healthy environment being reflected in environmental responsibilities which recognise both the broad spatial and temporal scales covered by sustainability. Central to this new citizenship is active participation in processes that lead to agreed outcomes which engender a sense of responsibility, especially for groups in society who have previously shown less involvement in such processes (Freeman et al. 1996). The involvement of the 'silent majority' is important, not least because they may include groups who can contribute significantly to sustainable development, both in social and environmental terms. As a basic example, research has demonstrated that those individuals least likely to make lifestyle changes to help the environment are young, male and politically inactive (Barr and Gilg in press). Accordingly, participation rests with involving, as a matter of priority, those who by definition are the least likely to be involved.

Achieving this goal will be extremely challenging and provides the focus for the remainder of this text, but within the specific context of LA21, two points of tension can be identified. First, as Freeman et al. (1996, 68) noted '… the key players are faced with the dilemma of where and how to position themselves in the process … "ownership" needs to include, where possible, those who are often under-represented in participation'. Existing structures for decision making in the 1990s (and to some extent today), were based on top-down processes with defined boundaries and clear working guidelines for specific groups. For example, public consultation was normally the result of public meetings or surveys. If a new style of democratic engagement was required, the logical end point of such a process might be that this engagement would result in outcomes that necessitated new forms of negotiation to emerge.

Second, Selman and Parker's (1999) study of LA21s in Gloucestershire, Lancashire, Leicester and Reading through in-depth interviews revealed the alternative interpretations and discourses of sustainable development evident between key actors. Apart from the challenges of new participatory processes highlighted by Freeman et al. (1996), Selman and Parker (1999) found that LA21 resulted in a series of 'storylines' dependent on the group concerned:

- LA21 coordinators felt that LA21 was ground-breaking, novel and demonstrated an idealistic view of sustainable development;
- elected members argued that LA21 assisted with efforts to strengthen local democracy but were sceptical concerning the magnitude of the task, deliverability and the overall motives of LA21 activities;
- officers welcomed LA21 as a fresh idea, but argued for a focus on delivering targets;
- stakeholders felt uncertain and unsure what their role was.

Overall, Selman and Parker (1999) argued that two dominant themes emerged from their work. First, the success of LA21 in generating new participants within the development of plans was questionable: 'LA21 may have attracted some remarkably dedicated volunteers, but it has a long way to go before the hearts and minds of the majority are won' (Selman and Parker 1999, 59). This level of participation, from the most dedicated and motivated represented a major problem with the LA21 process overall. The 'recycling' of environmental campaigners from the 1980s into the sustainability advocates of the 1990s raised questions over the viability of LA21 as a process to engage the community. A second point to emerge from Selman and Parker's (1999) work was the inherent conflict they identified between ecologically committed individuals (such as LA21 coordinators and members of the public) and the dominant theme of ecological modernisation advocated by elected members and officers in the local authorities. This split between what Giddings et al. (1996) has termed the 'material' and 'political' realities is evident through sustainable development, but is felt acutely within the context of localised debates relating to development.

Conclusion: From Government to Governance

The policy shifts outlined in this chapter have dealt with the overall strategic framework of implementing sustainable development within the UK. Yet the underlying theme relevant to this text pertains to the identifiable shift in policy from 'government' to 'governance' (Connelly and Smith 2003). The early responses to the sustainable development agenda emphasised the role of government and business in leading the way towards a sustainable economy. Subsequent changes in both government and wider social trends have steadily replaced the role of 'government' as a centralising and controlling body involving a small network of civil servants and stakeholders, into a facilitator of 'governance'. Governance involves the inclusion of a wide range of stakeholders in the process of decision making, devolved to appropriate spatial scales. The onset of regionalisation in UK administration and the formalisation of initiatives such as Community Strategies have cemented the potential role of local communities in decision making.

Hoverer, the move from government to governance also reflects a more specific point related to the role individuals have within governance for sustainability. As we have seen the role of individuals has progressively been increased to the point where the 2005 Sustainable Development Strategy (DEFRA 2005) is predicated on changing individual behaviours to achieve sustainable development. The negotiation of the process that resulted in this position is important to recognise, because to present it as a linear model would be misleading. As this chapter has made clear, the importance of individuals has been recognised at the local level since the Rio Earth Summit. Yet in central government, this recognition has been slow to emerge.

The major conclusions we can draw from this analysis of sustainable development policy at the four major institutional scales of administration relate to three major points that take into account both temporal and spatial processes of policy development. First, sustainable development has moved from a policy initiative to an over-arching strategic framework for future development and has therefore been 'mainstreamed'. All government departments now have sustainable development strategies and sustainability pervades all tiers of government in the UK. Second, sustainable development is now recognised as posing significant challenges to existing economic and social practices and is promoted as an agenda for 'change'. The emphasis in the 2005 Sustainable Development Strategy (DEFRA 2005) on 'limits' presents significant challenges for adjusting our economy towards sustainable production and consumption. Third, progress towards sustainable development will be made only when lifestyle and behaviour changes are reflected across a majority of the population. Currently, our lifestyles and daily practices reflect the economic and social systems in which we act. These are high-consumption and high-waste systems, focused around wealth creation. Although options exist for more sustainable lifestyles, the uptake of these is slow and still regarded as a marginal choice, not a desirable and mainstream option. Central government is rapidly appreciating that whilst the first has now been achieved and the second may partly be dealt with using economic and other instruments, the last challenge represents the most stubborn barrier in progressing towards sustainability. It is to the issue of behaviour change we now turn, first by an examination of our existing knowledge base related to behaviour change and then to an examination of the means by which it is possible to understand behaviour change in terms of people's everyday practices, specific lifestyle groups and the motivators and barriers that impact on these different types of activity and lifestyles.

PART 2
Perspectives

Chapter 4

Behaviour Change: Policy and Practice

Introduction: The 'Change' Agenda

As we saw in Chapter 3, sustainability policy in the United Kingdom has progressively shifted from a centralised 'top-down' towards a distributed 'bottom-up' approach to implementing sustainable development policy. In tandem, emphasis has moved from using economic instruments or a reliance on market forces to deliver sustainability, towards a realisation that behaviour change is necessary amongst the wider population if the goals of sustainable development are to be achieved. As Chapter 3 highlighted, initial conceptions of public involvement in sustainability centred around consultation, but the placement of the behaviour change agenda at the centre of the most recent Sustainable Development Strategy (DEFRA 2005) has reinforced the significance of the individual to sustainable development. As the National Consumer Council of Britain (NCC/NEF 2004, 4) has recently noted '… without the commitment and involvement of individuals, there is no chance of success'. The role of individuals within sustainable development has been conceptualised in different ways and this connects to the shifts cited above, from a technocratic view of change towards a greater focus on the use of democracy and deliberation. These shifts inevitably govern the interpretation and definition of what constitutes 'behaviour' for sustainable development (Jackson 2005). The multiple discourses surrounding sustainability ensure that successive political and academic definitions of sustainable behaviour relate to the three main domains of sustainability: environment, economy, society. This inevitably leads to contradictions that problematise the notion of 'being' sustainable. For example, the sustainable activity of 'buying local' may imply significant economic benefits, but it may also conflict with social and ethical agendas to support fairer trade from developing nations. Indeed, buying locally or from developing nations does not guarantee environmental sustainability. Consequently, discourses of sustainable consumption are both contested and in conflict (Seyfang 2003). This text does not seek to engage with the debates surrounding definitions, but rather seeks to set a practical agenda for one specific type of behaviour for sustainability. Environmental behaviour has received by far the most attention both conceptually and practically and presents a reasonably concise set of activities aimed at reducing environmental impacts of human activities. It is readily acknowledged here that the environmental action does cross a range of boundaries and cannot be isolated from other forms of sustainable behaviours, but environmental action does nonetheless represent a clear intent, which is often less evident with other forms of behaviour for sustainable development. We shall return to the issue of how environmental action is defined in Chapter 6, but at present a working definition is that environmental behaviour is any activity which is explicitly undertaken to obtain a net benefit for the natural environment.

This chapter will therefore plot the shifting discourses of citizen engagement in sustainable development through the use of two approaches. First, within the policy context, four case studies of behaviour change policies will be examined. Second, a reflection on these approaches will be undertaken with reference to the academic debates surrounding the engagement of citizens in sustainability.

In the first instance, the initial two case studies of behaviour change policy will be examined emanate from the Conservative administration's perspective on sustainable development. Grounded fundamentally in a technocentric view of sustainability, with a clear emphasis on reinforcing but not changing existing environmental policy, we will examine how two behaviour-change programmes (*Going for Green* and *Helping the Earth Begins at Home*) represented a discourses that were aligned to an alternative and 'green' lifestyle concept. The third case study (*Are You Doing Your Bit?*) developed by Labour in the late 1990s will then be explored, demonstrating the emerging role for sustainable forms of consumption in the behaviour change agenda. Finally, the current policy initiatives for behaviour change in the Sustainable Development Strategy (DEFRA 2005) will be examined. These represent a step-change in how behaviour is conceptualised and the means by which change can be achieved. Not least, this last case study will demonstrate how policy-makers have begun to recognise that an established principle of behaviour change (Awareness – Information – Decision – Action or A-I-D-A) cannot be relied upon to effect change.

The second part of the chapter seeks to examine how academics from the disciplines of geography and environmental psychology have problematised policy approaches to behaviour change. In the first instance, contemporary work within geography on environmental consciousness (e.g. Owens 2000) will be examined. The current epistemological frameworks within geography will be explored, highlighting the criticisms many geographers have levelled at 'rationalistic' conceptualisations of behaviour change, as demonstrated by the A-I-D-A approach cited above. The call by geographers for a 'deliberative-civic' perspective to change is examined and then critiqued within the context of alternative work by geographers such as Barr and Gilg (in press), Burton (2004) and Wilson (1997). These geographers have advocated the use of work from the sub-discipline of environmental psychology to appreciate an understanding of behaviour change. The chapter concludes by advocating an approach based on existing and reliable social-psychological frameworks of behaviour and behaviour change as a vehicle for investigating how more sustainable practices can be developed in society.

Engaging the Citizen: Behaviour Change Policy in the UK

This section examines four sequential approaches initiated by central government in the UK for changing behaviour towards the environment. As noted above, these approaches have each represented a particular perspective on both sustainable development and the way in which individuals can play a role in moving towards sustainability. Through a case study approach, this section will explore the discourses of behaviour change exemplified by these four policies.

Going for Green

The Conservative administration's initiative to engage citizens in environmental action for sustainable development was launched by the Department of the Environment a year after the publication of their Sustainable Development Strategy (DoE 1994). *Going for Green (GFG)* worked in partnership between central government and the private sector and was launched as a limited company. This ensured that it could raise funds from private sources and more easily engage in sponsorship deals within the consumer environment. Within this context, funding for *GFG* was obtained from major organisations such as Tesco, McDonalds, Hoover, Biffa and Kelloggs. *GFG* had five central messages that it wished to drive home to consumers, to be reflected in their behavioural commitment:

- travel sensibly;
- prevent pollution;
- cut down waste;
- save energy and natural resources;
- look after the local environment.

Through the use of these central behavioural messages, Collins (2004) argued that *GFG* made two key assumptions regarding behavioural change and public commitment. First, there was broad agreement that the 'problem'; of public participation was one of awareness. Such awareness included both a recognition by individuals of the problems posed by unsustainable forms of behaviour and also the potential for the individual citizen to overcome such difficulties. In other words, information was the key to behavioural change. Second, Collins (2004) argued that the campaign assumed that on the basis of this information, individuals would respond in similar ways to the messages provided. To this end, the broad campaign slogan of 'cut down on waste' would logically be interpreted as the need to change one's own behaviour in terms of reducing waste, reusing materials and recycling household goods.

Collins' (2004) analysis of *GFG* highlighted three major characteristics of the initiative of relevance to this text. First, *GFG* represented a 'top-down' approach to participation in environmental behaviour. *GFG* was implemented through the mode of local action groups (LAG's), who used the five-point 'green code' as a means by which to encourage behaviour change. To this end, the boundaries of environmental action had been pre-set for individuals and behaviour was framed within these limits. Second, *GFG* represented what Collins termed an (2004, 207) '… anthropocentric "light green" approach' to sustainable development. The campaign effectively repackaged long-standing environmental issues, which did not seek to tackle wider issues of sustainable development related to the more fundamental relationship between consumption and environment. Accordingly, Collins (2004) argued that many households found this irrelevant. Finally, as mentioned above, *GFG* represented a simplistic view of behaviour change, lacking theoretical rigour and omitting to acknowledge the role of different social and community groups and the different types of responses individuals within these groups were likely to have.

Going for Green therefore epitomised the Conservative government's approach to sustainable development that was stated in their Strategy (DoE 1994). It was an anthropocentric, weak view of sustainable development, highly compartmentalised around environmental issues and assumed a behaviour-change model that made the implicit link between awareness, information and action. However, *Going for Green* was a general environmental campaign. Alongside *GFG* ran a more specific initiative, *Helping the Earth Begins at Home*, to which we shall now turn.

Helping the Earth Begins at Home

Initially launched as a joint venture between the Department of Energy and Department of the Environment in 1991, this campaign was taken over by the DoE in 1992 as a result of the Department of Energy's closure. The campaign was a focused initiative that sought to make explicit links between personal energy use in the home and the impact of the carbon emissions from this usage on the global environment, most notably the greenhouse effect. Accordingly, campaign literature highlighted two key elements that linked energy consumption in the home to environmental problems. First, a representation of the Earth as a blazing sun highlighted the potential for global warming. Second, the image of a globe emerging from a chimney emphasised the link between personal energy consumption and global environmental issues (Hinchliffe 1996).

Helping the Earth Begins at Home was yet a further, but more specific, example of the anthropocentric, weak sustainability and 'top-down' approach to behaviour change. Hinchliffe (1996, 57) has highlighted the dominant anthropocentric discourses that were embedded in the campaign, highlighting the 'Conservative government's celebration of market forces' in the campaign messages, which emphasised both the benefits from saving energy in the home to be focused around saving money and the removal of central government from responsibility for regulating energy use. The central campaigning message was also representative of another 'top-down' approach, seeking a set of similar behavioural responses from the entire population, based on a weak link between information and behaviour, again highlighting (Hinchliffe 1996, 57) the role of government to '... encourage people to behave rationally at the micro scale and gain from the incidental benefits at the macro scale'. Accordingly, like *Going for Green*, this campaign was dominated by discourses of weak sustainability, compartmentalised environmental issues apparently unrelated to sustainable development and a generally weak 'top-down' approach to behaviour change, represented by aloof campaign messages outside of the mainstream social networks of the majority of the population.

Are You Doing Your Bit?

The Labour government's publication of their Sustainable Development Strategy (DETR, 1999b) witnessed a shift in emphasis in relation to the role of individual citizens and the definitions applied to sustainable development (see Chapter 3). These conceptual shifts were partially represented in the behaviour change policies that Labour instigated. The most prolific and high profile was the *Are You Doing Your Bit?* Campaign (AYDB) launched in 1999 (DETR 2000b). AYDB was once again a national campaign and awareness initiative that sought to encourage environmental action. Indeed, as previous Conservative initiatives has emphasised, key behavioural changes were highlighted by the messages and these were focused on small, incremental changes in the home. However, what defined and differentiated AYDB from previous campaigning styles and messages was the types of behaviours being advocated and the style of marketing employed. In the first instance, the broadened definition of sustainable development utilised by the Labour administration was reflected in the types of activities advocated by AYDB. For example, for the first time, sustainable consumption was highlighted as an activity worthy of consideration. Indeed, AYDB recommended very specific actions, quite different from the *Going for Green* initiative, such as turning off lights in unused rooms, travelling by public transport rather than private car, etc. However, a second shift represented a more fundamental change in approach to central government campaigns towards the environment. The use of multimedia campaigning meant that coverage of AYDB was broad, covering newspaper advertising, TV and radio broadcasts. These were short, snappy and had clear and concise messages, which were delivered by well-known celebrities. Indeed, messages were packaged in a more manageable fashion, emphasising incremental change to lifestyles. For example, campaign messages focused on types of behaviour change each month, such as energy, water or waste. Campaign 'road shows' also toured the country to provide more practical 'hands-on' information to consumers.

The AYDB campaign certainly represented a change in campaign delivery and to some extent content. It was without doubt a more mainstream initiative than what had gone before, being adopted by numerous organisations as a brand. Indeed, the emphasis of 'doing your bit' was the most explicit reference to citizen involvement in a wider social movement that had been made, in an attempt to emphasise the community and social dimension to personal behaviour change. Tied with new messages that made more ambitious proposals for changes to consumer lifestyles, this ensured that AYDB is probably the highest profile environmental campaign ever run in the UK. However, AYDB's surface changes disguised some crucial underlying assumptions that represented the mainstream discourse of behaviour change campaigning. First, there was a persistent belief that awareness and information was central to changing behaviour. To this end, a second key assumption was that such behaviour change campaigns at a national level would be effective almost universally. Finally, a central assumption in all three of the campaigns we have examined thus far is the notion that an individual's environmental consciousness enables them to transfer environmental messages into everyday lifestyles with ease. These three assumptions form the basis of a critique of policy that has been vigorously pursued within geographical literature. We will shortly turn to this critique, however it is pertinent to firstly examine the approach taken towards behaviour change in the current Sustainable Development Strategy (DEFRA 2005).

Community Action 2020 – Together We Can

DEFRA's (2005, 32) latest thinking on behaviour change has been informed by a range of research that has highlighted the complexities of behaviour change (Demos/Green Alliance 2003; Darnton 2004a; 2004b; Jackson 2005; Cabinet Office 2004) to the extent that it has admitted 'Evaluation of previous campaigns suggests that they have raised awareness but not translated into action'. Community Acton 2020 seeks to build on this work by creating capacity for behaviour change that uses an approach based on four key dimensions (Figure 4.1). Specifically for Community Action 2020, the government envisages:

- positive and inspirational messages rather than fear or concern;
- avoiding what it terms 'above the line' advertising on TV or billboards (effectively moving away from actions which the majority of the population are either unable or unlikely to engage with);
- the use of local communication networks;
- high profile national communications;
- a new environmental brand to link across communications.

An assessment of this initiative is too early to investigate, but there is evidently a shift from generalistic advertising towards more localised and specific campaigning. However, the detail currently available does not permit any more than the brief conclusion that government is slowly recognising the complexities of behaviour change.

Critiquing Environmental Policy and Approaches to Behaviour Change

The criticisms levelled at behaviour change policies have come from many quarters, including both policy analysts and academics. However, some of the most vigorous concerns have been raised by geographers who have based their critique on the fundamental assumptions made by those constructing awareness campaigns. This section will detail these critiques focussing in the first instance on key criticisms of current behaviour change policy, before opening out the debate to consider the implications that these criticisms have for finding a more effective set of policies for behaviour change.

The dominant research agenda within geography that has led the critique of behaviour change policy in Britain represents the 'cultural turn' in geographical scholarship. The manifestation of this shift has been the emergence of what, very broadly, can be termed culturally informed approaches to the study of various phenomena. As interest has grown within the discipline concerning environmental

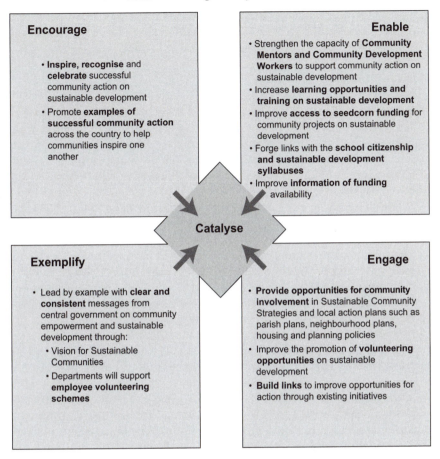

Figure 4.1   Community Action 2020 (adapted from DEFRA 2005). © Crown

Source: DEFRA (2005).

issues, often at the local level, there has been an integration of culturally-informed qualitative methodologies into work on environmental action and environmental policy.

   The basis of this approach is essentially an active deconstruction of the underlying assumptions about the social and political discourses that formulate the basis for environmental policy. This takes a number of forms, which are discussed in turn below, but in general, work into environmental policy and behaviour has focused on the 'scientisation' and 'politicisation' of the environmental debate, which has characterised environmental problems and their solutions by elitist, 'top-down' approaches, in contrast to the belief by many geographers that a more deliberative perspective on behaviour change is required. Work has also focused around the critique of policy frameworks that categorise and make assumptions regarding individuals as 'rational agents'. In other words, there is an implicit logic in policy which assumes knowledge of environmental issues will lead to concern and then action. We will now briefly examine these two arguments before examining the nature of the geographical critique in more detail.

A Deliberative and 'Civic' Approach

To researchers advocating a deliberative and 'civic' approach, the very notion of what constitutes environmental action is problematised within a wider political discourse that has become disconnected with society more generally. Predetermined actions set by national governments and promoted as being 'sustainable' are therefore constructed in ways that are not reflected in everyday social and environmental concerns.

Within this context, work in geography on effecting behavioural change has focused around what has been termed a 'deliberative' model of public engagement with sustainability (Owens 2000, 1141) and the ways in which individuals receive, interpret and act on environmental information in a range of discursive and institutional contexts. The deliberative model also proposes that engagement of the wider public will be forthcoming only when social and environmental problems are framed and debated within the spatial and temporal scales at which individuals are expected to take action (Blake 1999). Accordingly, (Hobson 2002, 113) argues that environmental action based on '... voluntary information and lifestyle initiatives will constantly create "discursive traps" ... by information presented in impersonal media'. Participation in action for sustainable development is therefore seen as contingent on a range of factors, relating to the nature of the information provided and its interpretation (Myers and Macnaghten 1998), based upon the trust with which it can be handled (Macnaghten and Jacobs 1997; Hobson 2001) and the complexity of scientific information (Eden 1996; Macnaghten and Urry 1998) provided by 'experts'. Overall, as Owens (2000, 1141–2) has noted, this 'civic' approach to examining public engagement with sustainability '... is less prescriptive of information flow and admits a wider range of understandings into the category of 'expertise' [such that] '... what is sought here is democratic engagement ... moving beyond the prescribed responses to predefined problems'. This democratic engagement has the aim not merely of identifying solutions to problems, but also to 'reframe' the nature of these challenges.

Rationality and Individual Action

A second major criticism of contemporary sustainability policy and a key argument for rethinking how individuals act is the notion that people are not, by their very nature, rational beings and that policy should reflect this. Burgess et al. (1998) reflect that 'sustainability is predicated on the belief that individuals and institutions can be persuaded to accept responsibility for the production of environmental problems and change their everyday practices to alleviate future impacts' (1446). This ultimately rests with what Macnaghten and Urry (1998) have termed 'people's ignorance of the facts' (212). In turn, enlightenment will lead to concern and hence action. Enlightenment evidently comes from the provision of environmental knowledge, in what Eden (1998, 426) terms 'the predominant 'linear model' of policy influence which assumes a one-way flow of information – from science to policy and society'. This rational, or 'objective' (Macnaghten and Jacobs 1997) model is how this critique characterises both policy and the research on which this policy is based. Attention now turns to the detail of how a set of assumptions can be used to understand environmental behaviour.

The Geographical 'Framing' of Environmental Action

The general criticisms of environmental policy highlighted by geographers (the 'top-down' approach and 'rational choice' model) have been used within the context of the search for a deliberative and civic approach to explore the notion of environmental citizenship. Accordingly, geographers have framed the behaviour change 'problem' through an analysis of three specific conceptual barriers to change. First, researchers have explored *political scepticism* with the implementation of environmental policy, focussing on the growing divide between science and society and the increased mistrust of scientifically-based policies. Second, researchers have found that the 'localisation' of environmental issues (Eden 2000), or the 'relatedness' of environmental issues to everyday life, can have significant impacts on the perception and response to policy measures aimed at focusing on global issues (e.g. Macnaghten and Jacobs 1997). This has been researched by a number of authors using discourse analysis (Hajer 2003). Third, workers have found that there are issues regarding the 'efficacy' of behaviour; individuals may well perceive personal responsibility for acting, but do not perceive that the scale of global environmental problems can be aided by their 'small' actions (e.g. Eden 1993).

Political Scepticism: 'Scientisation' and 'Politicisation' of the Environmental Debate

The first significant barrier to participation identified by geographers researching environmental action has been that the environmental debate represents an elitist discourse which in turn reduces public participation through the process of disenfranchisement. Eden (1996; 1998) has argued that scientisation occurs because a greater number of environmental problems are those which '... are increasingly only evident through science's sensory organs, rather than people's everyday senses' (Eden 1996, 187). This 'scientisation' process relates directly to the third of Beck's (1992; 1999) constituents of contemporary 'risk society', whereby environmental risks are only 'sensed' by scientific knowledge. In this way, science frames and produces highly specific information and presents a unique epistemology on environmental problems. Bulkely (2001) illustrates this with reference to global warming, arguing that the highly scientised nature of debates surrounding greenhouse gas emissions, framed within an elitist discourse, presents an environmental issue that cannot be sensed or debated through lay discourses. Eden (1998, 437) takes the debate further, arguing that the traditional notion of science as set apart from the policy-making process must be drawn into question given the scientific basis for almost all current environmental policy, such that '... it is now impossible for science to remain "aloof" from the policy process' (Eden 1998, 437). Yet this politicisation of science, as will be seen below, does not only raise questions regarding the role of science in producing and interpreting environmental issues, but also with regard to reflexivity (see the conceptual model of Beck, 1992). Certain environmental crises have focused attention on these issues, such as the BSE affair and more recently, the Foot and Mouth epidemic. Science has been 'found wanting', with 'experts' subsequently being proven to have misrepresented or misinterpreted information. Eden (1996; 1998) characterises the underlying assumptions of this argument by describing this as the 'expertisation' of science, leading to what she terms the 'self-censorship' (1996, 191) of ordinary peoples' critique of science. Because of this self-censorship and general mistrust of science and government, people seek other forms of knowledge and understanding about the environment. Eden (1998) provides examples of how some policy-makers have attempted to fill the democratic deficit in science by incorporating, in some way, other knowledge, for example locally contextual understandings and even '...emotional, vernacular and moral input from nonscientific publics, particularly with ties to local knowledge' (Eden 1998, 429). These latter suggestions are intended to increase environmental action by (re)enfranchising lay publics.

More specifically, the importance of 'trust' was an issue further evidenced from focus group discussions by Macnaghten and Jacobs (1997) who argued that political mistrust is a significant barrier to providing a basis for individual responsibility and efficacy. Central government was seen as distrustful, corrupt, hypocritical and unaccountable by their focus group participants. There was a feeling that the 'political system' was to blame, encapsulated by self-interest and personal gain. However, Macnaghten and Jacobs (1997) and Macnaghten and Urry (1998) show how this apathy also encompasses other groups who may purvey untrustworthy information, such as academics and bureaucrats, such that, as mentioned above, knowledge of environmental issues was limited to immediately sensed phenomena. In the extreme, the '... role of the state within the official discourses of sustainability was apparently rejected. Institutions of the state were generally seen to be part of "the system" which generates environmental and social problems ...' (Macnaghten and Urry 1998, 231).

Local Environmental Discourses

A second barrier to environmental action identified by geographers is the way in which individuals conceptualise environmental issues and the means which they use to understand such problems. Burningham and O'Brien (1994, 929) have argued that '... frameworks for environmental understanding and action cannot be imposed from outside such contexts but must be generated from within through attention to the different ways that values and motives are localised'. Macnaghten and Jacobs (1997) and Macnaghten and Urry (1998) demonstrate this process by showing how the general public identify with sustainable development. They argue that the rhetoric and language of sustainable development can be seen in terms of localised interpretations of phrases such as 'quality of life', which Macnaghten and Jacobs (1997, 13) describe as reflecting mainly '... personal, individual concerns', focused around employment, crime and local environmental quality. Issues of environmental quality were focused primarily on local problems, such as dog mess, car pollution or development issues. Indeed, Macnaghten and Jacobs (1997, 16) argue that so focused were their young male focus group on their own lives and the prospects therein, that '... the very concept of "the environment" appeared so abstract and removed from their everyday lives that discussion was significantly impeded'. Macnaghten and Urry (1998) develop these themes further and argue that environmental issues are bound up in the way each individual lives his or her life, such that people referred to the impact of environmental change '... in daily practice' but also to '... dangers that were distant, unseen, unknown, or delayed' (235).

Overwhelmingly though, 'environmental discourse' (Hajer 1995) at the individual level appears to be rooted in local and contextual discursive narratives that frame environmental issues in everyday life. Examples of this approach pertinent to environmental action include Petts' (1995) study of waste management planning in Hampshire and Goodwin's (1998) study of conservation. This is not to state that issues of global significance are either ignored or unknown. Rather, issues at the local level can be more readily sensed and, overall, might be resolved efficaciously.

The role that a range of local discourses can play is demonstrated by Linnros and Hallin (2001) in their study of the Oresund road link between Denmark and Sweden. Using the notion of a 'core-periphery' model of discourse, they argued that the dominant 'regional evolution' discourse, those promoting the road link and supported by national and regional government, were able to occupy the core area of what the authors termed 'self evident and unquestionable' (392) statements, whereas the three major interest groups were on the periphery. Through using this discursive core, the proponents of the link were able to successfully utilise discursive agents (e.g. policy-makers, planners, journalists) in an appropriate discursive arena (e.g. the media) to adopt a discursive strategy (exclusion of other arguments) based on the spatial representation of a modern utopia which the link would bring.

Response Efficacy and Environmental Responsibility

Eden (1993) has provided a coherent account of the issues pertaining to environmental responsibility. In particular, she acknowledges that response efficacy and responsibility are intricately linked in the relations between an individual feeling there is a need for action and actually prescribing responsibility for acting to the self (Schwartz 1977). This distinction is important, since in Eden's (1993) study of environmental activists, she found that '... a lack of responsible agents [i.e. the public] would seem to strengthen the individual's [i.e. an activist] perception of environmental responsibility, but only where efficacy is strongly perceived' (1755). A number of articles have charted the nature and structure of response efficacy and responsibility. Macnaghten and Jacobs (1997) and Macnaghten and Urry (1998) have shown how their focus groups demonstrated a willingness to act and some genuine acceptance of responsibility for behaviour, but that this might only be achieved efficaciously at the local level because, effectively, they were powerless to act. The authors relate this to the mistrust of political and scientific institutions as discussed in the previous sections and shows the trend outlined by Eden (1993) for issues of efficacy to form a barrier between awareness of need for action and the ascription of responsibility to the self.

Hinchliffe (1996) provides further evidence of this phenomenon with his analysis of the Conservative government's *Helping the Earth Begins at Home*. Hinchliffe argues that the focus on one, abstract and ultimately debatable environmental issue (global warming) provided the groundwork for scepticism about individual responsibility, encompassed again by notions of a lack of response efficacy. Hinchliffe (1996, 59) states that '... the scope for *individual* householders to make significant or even slight changes to their lifestyles ... was regarded as extremely limited'. Ultimately, this led to the conclusion by respondents that energy conservation was '... the responsibility of distant and equally abstract institutions' (61). Harrison et al. (1996) provide more support for this thesis regarding environmental responsibility. In their group discussions on environmental action, Harrison et al. compared groups of individuals from Eindhoven in the Netherlands and Nottingham in England. They allude to an erosion of responsibility and 'collectivism' in the English groups which is put down to the alienation and frustration felt by participants in that their actions towards the environment were not efficacious because of the lack of moral acceptance of the behaviours. This was ascribed to the social and moral decline emanating from the 1960s, the growth of the consumer society, political scandals and so on. In contrast, the 'collectivism' demonstrated in the Netherlands was perceived to be a source of greater environmental responsibility.

These studies, primarily drawn from the geographical literature, propose one epistemological framework that has sought to focus on the cultural and institutional contexts within which behaviour is framed. Behaviour change is therefore seen as dependant on deliberative and inclusionary processes. In contrast, evidence suggests that policy continues to approach the issues of environmental action and behaviour change from what Hobson (2002) terms a 'rationalistic' perspective, emphasising knowledge and awareness of specific environmental problems as the mode by which to effect change and close the 'value-action' gap. In policy terms, environmental action is framed by existing environmental issues that confront the state, such as an 'energy' or 'water' crises. The determination of how such issues can be addressed is presented to citizens both at varying spatial and temporal scales, through the mode of information transfer which encourages individuals to 'do their bit'.

Geography and Policy

As we have seen in the previous two sections, two distinct approaches and perspectives on behaviour change can be identified and characterised relating to:

- an *information-intensive* approach, advocated by policy-makers, grounded in key assumptions regarding the nature of behaviour change and the belief that raising public awareness will result in a 'linear' transformation from attitudes to behaviour;
- a *deliberative* approach, advocated by geographers, which places emphasis on the contested meanings of environmental behaviours and the institutional and social barriers to behaviour change, arguing for a deeper 'civic' approach to policy-making.

Epistemologically, these approaches represent alternative perspectives on how environmental knowledge is produced, disseminated and interpreted. In terms of knowledge production and interpretation, policy approaches have conventionally argued for a 'rational choice' model, whereas geographers have argued that attempting to understand behaviour in terms of a series of rational decisions is highly problematic. Indeed, policy-makers have conventionally utilised an approach that stipulates targets and attempts to shift behaviours accordingly. In contrast, geographers have argued for a deliberative model that empowers communities to define and frame environmental problems as a means by which to encourage behaviour change.

The contrast in these two perspectives could therefore be examined in purely epistemological terms, but the differences also reflect methodological issues, most crucially concerning how environmental action is framed, interpreted, explained and changed. These differences have been highlighted by researchers such as Owens (2000) who have critiqued the notion of rationalising environmental practices through the use of 'models', effectively restricting the citizen to a predetermined set of influencing factors. The argument does, in many cases (e.g. Macnaghten and Urry 1998) proceed further to a more detailed critique of the methodological framework used by researchers for attempting to understand behaviour using quantitative data. Such a methodological approach has been the basis for considerable research on environmental action by social-psychologists and has partly been the focus for criticism by geographers in their wider critique of 'rationalistic' approach to behaviour change. As will be seen later in this chapter and in Chapter 5, the social-psychological approach promotes a more quantitative framework for exploring behaviour change and does use modelling for its work. However, despite the large amount of research in psychology that has explored environmental behaviour, there has been reluctance amongst many geographers to engage more fully with this work apart from the general criticisms discussed above. Indeed, given the disciplinary boundaries within the UK, there has been little incentive for interdisciplinary working on this issue and from the perspective of geography therefore, a research agenda has emerged which has rightly critiqued the dominant political discourse on behaviour change, but has rarely engaged with academics from other disciplines who research this area.

Accordingly, a new perspective on environmental action calls for a reappraisal of both epistemological and methodological assumptions, most notably the concerns of geographers that simply making individuals aware of environmental issues will elicit behavioural changes. From the perspective of psychology, it is certainly the case that the development of social-psychological frameworks has examined the impact of knowledge/awareness on action using quantitative methods. However, to make a simple link between such research and the potential dismissal of a wider methodological approach is problematic. Indeed, social-psychologists have provided critiques of current policy for behaviour change with equal vigour to that of geographers. Research into environmental behaviour by social-psychologists and quantitative sociologists has provided a wealth of material that needs to be explored with equal merit, epistemologically and methodologically (see the review in Chapter 5). This is an argument which has been advanced by geographers such as Wilson (1996; 1997), Burton (2004) and Burton and Wilson (2006) in their studies of farmer attitudes and behaviour. These geographers have adopted this alternative epistemological and methodological approach, based on a theoretically guided social-psychological perspective, in order to address the issue of behaviour change for sustainable development. Implicit in their assumptions is an appreciation that behaviour *is* complex, but that the means by which to understand behaviour and potentially to effect change is to provide policy-makers with information about what the nature and structure of environmental action is, what appears to influence it and who the dominant participants are within a quantitative framework. Using these assumptions, research has sought to conceptualise environmental action by studying policy-defined behaviours, but using an approach that problematises their definition and frames them within everyday practices. Indeed, such work seeks to understand environmental behaviour within the wider context of research that has examined a range of social behaviours using psychological frameworks to refine understandings of behaviour.

The Social-Psychological Perspective

The psychological approach to studying attitudes and behaviours has a long and diverse history. Thurstone and Chave (1929) are regarded as two of the pioneering scientists to advance the modern notion of attitude. The study of attitudes and behaviours was brought into sharp perspective by LaPiere's research in the 1930s. LaPiere (1934) noted that the overtly hostile verbally expressed attitudes of Americans towards Oriental descendants were not replicated by their behaviour towards Orientals. The disparity between attitudes and behaviours, or the 'discrepancy' (Eiser 1986) is still what forms the basis of social-psychological inquiry today.

A number of approaches to the social-psychological study of attitudes and behaviours can be identified and in particular can be grouped into two genres

(Barr 2002). The first is what might be termed the 'modelling' approach. The second can be termed the 'framework' approach. This categorisation has particular salience with regard to environmental behaviour and environmental psychology in general.

In the first instance, the 'modelling' approach is the more rigid and theoretically orthodox of the two. This approach seeks to define the relationships between given independent and dependent variables on the basis of their positioning within a given theoretical framework, the associations between the factors having been fixed. A good example would be the theoretical model of Schwartz (1977) which is a processual model of 'helping behaviour' (or, as it has become known, a model of altruistic behaviour). Schwartz posits a series of hierarchically defined steps from an initial stage of 'Awareness of need' to 'Action'. The model is rigid; iteration of steps is only permitted at certain stages and a given outcome is fixed according to antecedent variables.

This model has been successfully tested using, for example, recycling or 'yard burning' behaviour (Hopper and Nielsen 1991). However, the point to note about Schwartz's model is that in this form, variables cannot be adjusted, added, excluded and nor indeed can the relationship between them be altered. This is not to state that this is wrong in any way; for certain behaviours, this may be a very accurate means by which they can be understood. Rather, it demonstrates that the model is specific and fixed. Assumptions are made and relationships are rigid.

This can be contrasted to the second approach, which is what has been termed above the 'framework' approach. It is characterised by far less theoretical rigour than the modelling approach. It might even, in certain circumstances, be termed 'ad hoc'. There is considerable variation in this approach. Some studies have defined a framework of behaviour and then tested it as with a model (e.g. Guagnano et al. 1995). Others have examined the most likely predictors of a given environmental behaviour and used these to find the best subset of predictors by correlation methods (e.g. Steel 1996). In general, the framework approach is characterised by some theoretical, but mostly empirical, understanding of the behaviour concerned. This understanding is used to build a framework of behaviour. There is a significant difference between what has been termed a model and what has been termed a framework. Although on paper they may look similar, the means by which the researcher will manipulate the variables is important. First, no variables *have* to be included; their inclusion may depend on the empirical and other evidence. Second, the positioning of variables can change if their effect is seen as more important in another relationship. Third, the relationship between any two or more of the variables is not set; they can be changed and most importantly the framework amended after testing.

A good example of this approach is that used by Barr et al. (2001) in their study of waste management behaviour. This is explained in more detail below, but is based on a model (the Theory of Reasoned Action of Fishbein and Ajzen 1975), which has been loosened into a framework to take into account certain empirical circumstances. The framework was tested, relationships amended and finally altered to take account of the results. Hence, the framework did not 'pass' or 'fail', but rather it was used as part of a learning process.

In the context of this text, it is arguable that the more appropriate approach to use for examining environmental action is the framework methodology. To demonstrate how this is the case and to provide an insight into the social-psychological literature involved, the Theory of Reasoned Action (TRA), one prominent theory of social behaviour, will be discussed. Through analysis of the literature on environmental behaviour, it will be shown how this basic model can be used to construct a framework of environmental behaviour (Chapter 5).

The Theory of Reasoned Action and its Deficiencies

The Theory of Reasoned Action (TRA), as shown in Figure 4.2 (Fishbein 1967; Fishbein and Ajzen 1975) has been frequently applied to environmental behaviour. The theory posits that action, or the behaviour, is a direct result of stated intention to act. In the original conceptualisation, Fishbein and Ajzen (1975) argued that this was the only logical precursor to action. Intention was predicted by just two other factors: attitude and subjective norms, the former comprising the anticipated consequences of an action and the evaluation of these, the latter comprising the awareness and acceptance of social norms to act. Later, Ajzen (1991) argued that both intention and action could also be predicted by what was termed 'perceived behavioural control', or in other words the self-efficacy of acting. This was in part a response to the deficiencies of the rigid model, which are examined below. The adaptation was termed the 'Theory of Planned Behaviour' (TPB). However, what the TRA and TPB show is a good example of a rigid and processual model which is limited to certain predictor variables and which can be changed very little.

Research into environmental behaviour has and is still using the TRA and TPB. What this research shows with regard to environmental behaviour is that two problems consistently emerge when attempting to use the model to predict, for example, recycling behaviour. The first has to do with the limitations posed by the variables within the model. For example, Lam's (1999) study of water saving using the TPB found that the addition of variables measuring the perceived moral obligation to act as well as the perceived right to a supply of fresh water both predicted behavioural intention independently of attitude and subjective norms. Other authors have found that behavioural intention can be better predicted by other factors (e.g. Kok and Siero's analysis of tin can recycling, 1985 and Boldero's study of recycling, 1995).

Second, not only do different factors provide better predictions of behavioural intention, but authors have found that the relationship between intention and action can also be interrupted by other factors, such as behavioural experience (Macey and Brown's study of energy saving, 1983; Goldenhar and Connell's study of paper recycling, 1992–1993) and environmental knowledge (Kaiser et al.'s study of general ecological behaviour, 1999).

These two problems – identifying the predictors of behavioural intention and those of behaviour – show how it is not necessarily acceptable to use a model of behaviour to predict environmental action. Considerable variation is still likely to be unexplained even after more variables have been added. Indeed, the relationships between the variables are also likely to be unclear. What is therefore

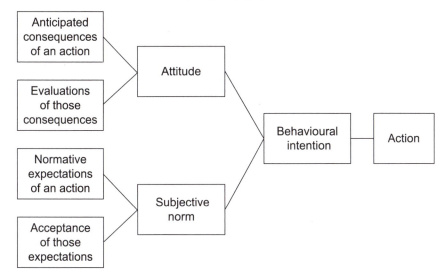

Figure 4.2   The Theory of Reasoned Action (based on Fishbein and Ajzen, 1975)

needed is a framework approach to assessing environmental action. This will be described in more detail in Chapter 5. However, in order to provide a coherent framework, it is necessary to understand more fully the work that researchers from social-psychology, sociology and geography have been undertaking in order to attempt to explain why people act in the way they do towards the environment.

Linking Policy and Research: A New Agenda for behaviour change

The assumptions underlying the social-psychological approach provide the basis, as detailed above, for a new perspective on the study of environmental action, attempting to provide middle ground between the culturally informed critique of current behaviour change agendas and the 'linear' and 'rationalistic' tools of current policy. It is argued here that this can be undertaken by focusing understandings of behaviour and behaviour change on a transparent framework of environmental action derived from the literature, utilised to deal with three specific policy-related problems that have emerged in recent years. Figure 4.3 provides an essential and basic framework for understanding environmental action. Based on the TRA (Fishbein and Ajzen 1975) and the Theory of Planned Behaviour (Ajzen 1991) this framework of environmental action seeks to identify barriers to behaviour change through the appreciation of the 'gap' that exists between intentions and action, so clearly recognised by the current Sustainable Development Strategy (DEFRA 2005). The framework posits that three sets of factors influence both intentions and behaviour, relating to *social and environmental* values, *situational* and *psychological* variables. Chapter 5 will examine these in greater detail, but essentially the logic of the framework implies that social and environmental values, as the guiding principles in everyday life, should determine our intentions and behaviour towards the environment. However, both situational characteristics and psychological variables intervene to modify this relationship. Situational variables include structural variables such as service provision and household circumstances, as well as socio-demographic composition and exposure to environmental knowledge. Psychological variables are personality and perceptual characteristics of the individual and include attitudes towards undertaking specific behaviours. Such variables can include perceptions of convenience, effort, responsibility, etc.

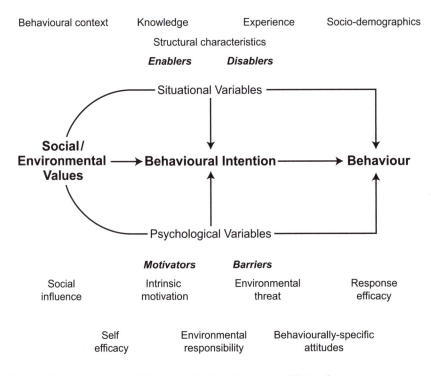

Figure 4.3   A conceptual framework of environmental behaviour

This framework can assist in the identification of key factors that determine levels of behavioural commitment. However, an analysis of past and current policy would indicate two further challenges that lie ahead. The framework outlined above and in Figure 4.3 assumes that behaviours are clearly defined and that the population can be treated as a whole when examining specific barriers. Yet campaigns such as *Are You Doing Your Bit?* have clearly demonstrated the compartmentalised nature of environmental action, stressing different behaviours related to specific environmental problems (such as energy, waste, water, etc.). It is doubtful whether the lived experiences of individuals conform to these compartmentalised notions of activity. Rather, it is necessary to examine how policy messages can relate more closely to the individual experiences of daily practices undertaken through the course of consumption of different resources. For example, retail consumption can involve multiple environmental behaviours, through reducing waste, saving energy or engaging in green forms of consumption depending of what is purchased.

This focus on daily practices also needs to be reflected by an acknowledgment that society, far from being a homogenised group of like-minded individuals, can be segmented into a series of groups. This common technique amongst market researchers is now being developed in the field of sustainability and the question that arises is the extent to which it is possible to identify particular groups of individuals with similar levels of environmental commitment. Current branding and marketing strategies focus on relationship marketing techniques that attempt to brand consumer goods at defined lifestyle groups. There is no reason why the logic of segmentation and the use of branding as a means by which to encourage environmental action cannot be examined to explore the extent to which lifestyle groups exist related to environmental practices.

As the foundation for an investigation of environmental action, based on current policy challenges, three approaches can therefore be identified that will be pursued through this text and which are grounded within the overall framework of environmental behaviour given in Figure 4.3:

- exploring environmental practices: understanding how environmental actions are related to each other and the daily practices in and around the home;
- identifying lifestyle groups: appreciating the potential lifestyle segments that may be evident in the population;
- examining barriers to and motivators for action: using lifestyle groups and our understanding of environmental practices to identify the possible ways in which behaviour can be encouraged.

This basic structure provides the context for the remaining parts of this book. The next task is to examine the role that specific motivators and barriers have played in providing an overarching framework of environmental action.

Chapter 5

The Social Psychology of Environmental Action

Environmental Behaviour research and the American legacy

The dominant discourse within contemporary British geography which has critiqued the 'rationalistic' model of behaviour change has developed largely independently from the growing body of research into what is more commonly referred to as 'Environmental Behaviour' (EB). EB research refers mainly to a series of approaches that use assumptions grounded in the social psychological subdiscipline examined in Chapter 4. Accordingly, EB research is driven by, but is by no means solely populated with, psychologists. The intellectual foci of EB researchers can be placed into two main groups relating to the epistemological and methodological assertions examined in Chapter 4:

- *epistemological*: using psychological principles as the basis of examining behavioural changes and the basis for this change;
- *methodological*: using a primarily quantitative approach, based on 'reliable' psychological measures to study behaviour change.

Epistemologically, EB research has a heritage within the broader discipline of psychology, out of which has developed a vibrant and growing subdiscipline which has become termed 'environmental psychology' (Canter 1987). This subdiscipline emerged in the 1970s as a reaction to the emerging importance of the 'environment' as a social concern. However, in this context, the word 'environment' is used in a wider context. Environmental psychology is concerned with issues as diverse as architectural design, office layout, home environments and recycling behaviour. As a subdiscipline, environmental psychology developed a distinctive set of working procedures and models that made it distinctive from other forms of social psychology.

Methodologically, EB research has taken many of the essential psychological assumptions underlying research practice and applied them to a diversity of contexts. The methodological perspective taken by psychologists has been driven by two dominant themes: scientific rigour and reliability and an incremental conceptual basis for this scientific enquiry. In the first instance and drawing from the natural sciences, EB research has employed the principles of scientific rigour. This can be seen within the context of both research design and analysis. Psychologists place great emphasis on being able to substantiate and justify a research design as being 'rigorous'. In practical terms, this implies that the survey design must be reliable; that is to say, if it were re-run under the same conditions, the results would be the same, within the given margin of error. The application of this 'reliability' approach also pertains to analysis, where the heavy reliance on statistical methods is designed to ensure both that the use of existing psychological measures can be tested in different circumstances, but also to ensure that readers of research outputs can compare findings across studies. Accordingly, this reliance on the scientific approach to research ensures that research findings are relatively simple to interpret, although an understanding of the various statistical terms is normally essential, given the large number of assumptions made by psychologists concerning the reader's knowledge.

EB research has therefore emerged as a distinctive subdiscipline of psychology, creating its own theoretical models and applying rigorous psychological testing to them. However, there are two issues that make environmental psychology research distinctive. First, the legacy of EB research has without doubt been influenced by progress in a range of distinctive geographical contexts. As Canter (1987) notes, the emergence of environmental psychology within the UK was relatively slow, yet EB research was well established by the early 1980s in the United States. With specific reference to environmentally positive behaviour (such as recycling, energy saving, etc.), work in the United States by the end of the 1970s (e.g. Dunlap and Van Liere 1978) was making significant contributions to the theoretical basis for an understanding of environmental behaviour. Other areas of the world that have seen advances in this field include Australasia and countries such as the Netherlands and Sweden. Yet it was not until 1999 that a preliminary meeting was convened of what has become 'Environmental Psychology in the UK' (EPUK), a forum for UK researchers. The UK has only one dedicated environmental psychology research unit (at the University of Surrey) and there is as yet no established lobbying group for those with interests in this area.

This lack of institutional focus can be seen as a disability for the interests of environmental psychologists. However, for the subdiscipline itself, it can be seen as a significant benefit. The first meeting of EPUK at London Guildhall University in 1999 attracted research workers from a range of non-psychology disciplines and policy institutes, including sociology, economics, geography and environmental science. Accordingly, the second distinctive feature of environmental psychology and EB research is its interdisciplinary nature. Although all of these researchers are utilising similar approaches and methods, their disciplinary background provides them with a unique perspective on issues that concern their own disciplines.

EB research therefore spans disciplinary boundaries but still maintains a focussed and scientifically rigorous approach to research. The following sections utilise this research to review a very small, but nonetheless highly significant area of EB work on the understanding and promotion of environmentally positive behaviours. Because of the scope of this text, the review will examine research that has examined four key environmental behaviours: energy saving, water conservation, household waste management and 'green consumption'. To set the context for this work, we will firstly examine the nature of these four apparently definitive activities

Contemporary Environmental Behaviour

Although contemporary environmental behaviour is contested and has been problematised by a range of authors (see Jackson 2005) the study of environmental action nonetheless requires a practical exploration of the types of activities which constitute environmental behaviours. The reader will undoubtedly be familiar with the seemingly limitless advice provided by a range of private and public organisations for becoming more 'environmentally friendly', from re-using plastic shopping bags to composting food and garden waste. It is not the intention in this book to dwell at length on any of these practices in detail, but rather to provide a synthesis of the types of activities which have broadly been termed 'environmental' in nature and the way in which they are framed and influenced.

As will be detailed in Chapter 6, the empirical research for this book started in 2001 with an Economic and Social Research Council (ESRC) grant exploring *Environmental Action in and Around the Home* (Barr et al. 2003). This research sought to link policy initiatives for behaviour change to reported behaviours and attitudes of individual citizens. However, the first major task for the research was to reflect on what constituted environmental action and how it could be measured. The research team decided to forge this link between research and policy by focussing on four types of activity promoted by the *Are You Doing Your Bit?* (DETR 2000b) initiative explored in Chapter 4, involving energy saving, water conservation, recycling and 'green consumption'. As the reader will note, sustainable travel is not part of this list, reflecting the significant academic and practical challenges with incorporating this field into an already large group of behaviours.[1] Accordingly, the remainder of this book focuses on a study of these four types of activity. The following four subsections explore how these different types of environmental behaviours have been defined and therefore provide the context for Chapters 6 to 9.

Energy Saving

Studies that have examined energy conservation have used a number of different labels to categorise this type of behaviour. However, an assessment of these studies shows that two fundamental categories emerge in most cases. What can be termed direct energy saving choices (Stern 1992a), or variably named 'adjustments' (Dillman et al. 1983), 'usage-related' (Van Raaij and Verhallen 1983) or 'curtailment' (Black et al. 1985) are focused around everyday reductions in energy use that require either no or minimal structural adjustment. Such behaviours include: thermostat setting, closing off of unused rooms, altering room use, window closure when heating is on, using a clothes line rather than a tumble drier, filling the kettle full before boiling, putting a full load of washing on rather than a half load. These choices also relate closely to those advised by the *Are You Doing Your Bit?* campaign.

---

1   The author has a further ESRC grant to explore these issues entitled 'Promoting Sustainable Travel: A Social Marketing Approach' running from 2007–2009.

This is in contrast somewhat to what Stern (1992a) has termed 'technology choices', also referred to as 'conserving actions' (Dillman et al. 1983), 'purchase-related behaviour' (Van Raaij and Verhallen 1983) and 'energy efficiency choices' (Black et al. 1985). These behaviours are often long-term alterations to the structure of the home building or at least changes internally that will require financial and normally technical resources to be utilised. This group of activities is more disparate than the first in the sense that the amount of financial and other resources can vary greatly, from, for example, full double-glazing installation to the insulation of a back door. Typically, behaviours measured in previous work have included: insulation (wall, door and roof), double glazing, purchase of energy saving products (appliances which are purchased with saving energy as a priority, such as washing machines, cookers, fires, dishwashers, etc.), using low-energy light bulbs and adjusting curtain heights in order to reduce energy loss.

Energy conservation activities evidently comprise a diverse group of actions in terms of both fiscal and personal commitments, but there is also complexity in terms of the problems raised in attempting to identify behaviours as definitively 'energy saving' actions. For example, it may be that a given individual purchases a highly energy efficient washing machine because there are other features of the product which appeal and not necessarily its energy efficiency. Indeed, it should also be noted that not all homes would be able to or would need to install, for example, roof insulation.

Water Conservation

As Berk et al. (1980) have noted, reducing the amount of water that is required at the micro (household/individual) level is the core concern of policy that seeks to achieve overall water conservation. In this sense, one could imagine a water hierarchy akin to the waste hierarchy, with 'water minimisation' measures (such as water meters and shower flow devices) being set above 'water re-use' measures, such as water butts in the garden. Such hypothecation is, however, somewhat unnecessary and also complex, since the majority of water saving measures are by definition minimisation actions.

Beyond this debate, Berk et al. (1993) and Lam (1999) have provided the most useful schema for categorising water saving actions, along the same lines as those used for energy saving. What Berk et al. (1993) refer to as technological fixes within the home, or what Lam (1999) refers to as efficiency measures can be described essentially as for energy behaviour as 'technology' actions. With regard to water saving, these can be described as follows (Berk et al. 1993; Corraliza and Berenguer 2000; Hutton and McNiell 1983; Lam 1999):

- having a water meter;
- using a water butt;
- having a water saving device in the toilet cistern;
- shower flow device;
- reduction in water pressure.

What Berk et al. (1993) term 'behavioural' actions and Lam (1999) refers to as curtailment behaviours can be described, as before, with recourse to the energy literature as 'direct' water saving actions. These include:

* reducing toilet flushes;
* reducing length and frequency of showers and baths;
* turning off the tap during cleaning teeth;
* not running the tap when washing dishes;
* turning off water when 'soaping up' in the shower;
* controlling the use of water in the garden.

The last of these behaviours provides a point of interest that is highlighted by Hutton and McNeill (1983) and also Syme et al. (1990–91) from an Australian context. Traditionally, water saving in the garden has been a major issue in countries like Australia, but it is notable that in the UK that recent drought events have witnessed a growing concern around water use in the garden and the promotion of drought-resistant plants.

Waste Management and Recycling

Coggins (1994) has argued for a recognition of waste management in relation to the Waste Hierarchy (DETR, 2000) that places most emphasis on waste reduction, re-use and finally recycling. Barr et al. (2001) have clearly demonstrated the role that these three types of behaviour have in effecting sustainable waste management and the extent to which they are distinctive behavioural activities. 'Waste reduction' refers to activities that prevent materials being used that may become part of the waste stream. This might include reducing the use of plastic bags when shopping, looking for produce with less packaging or purchasing products made from recycled materials. 'Re-use' is the utilisation of a product without involving any material change in its nature, such as re-using shopping bags, glass jars and bottles or plastic containers. 'Recycling' refers to the activity of mechanically reproducing a product or an alternative commodity from its existing state, such as paper or glass recycling.

Green Consumption

In contrast to the relatively simple definitions for the previous three behaviours, green consumption is a highly contested and diverse form of behaviour.

The SRS and socially conscious consumption The main indicator of socially conscious or green consumption used in the earliest studies of green consumerism (e.g. Anderson and Cunningham 1972; Webster 1975) was traditionally the 'Social Responsibility Scale' (SRS). The SRS was developed and tested by Berkowitz and Lutterman (1968) who contended that the scale represented one form of social responsibility, encapsulated by '... *traditional, conventional* responsibility' (173, original emphasis). Thus, other forms of socially responsible action, such as citizen protest at a military dictatorship, which go against the social norm, were not included in the SRS. The SRS examines a number of personality attributes and behaviours that Berkowitz and Lutterman considered to be representative of a traditionally socially responsible individual (in the early 1960s). The scale has the following underlying traits:

- belief in one's ability to change events (power or 'response efficacy');
- giving up personal time to 'do good';
- belief in democracy;
- keeping to your word;
- doing a job 'the very best one can';
- sense of community;
- volunteerism;
- finishing a job.

The SRS must evidently be seen in historical social context, being developed in the early 1960s. There are similarities to Etzioni's (1995) conceptualisation of communitarianism that is still popular amongst New Labour and the recent moves to promote 'neighbourliness' by the British Conservative Party. However, for the purposes of this review, the fundamental point is that at the time the SRS was seen as a means by which to associate personality with responsible purchasing behaviour and was seen as the major proxy of such behaviour. The use of the SRS as a surrogate for ecological purchasing was short-lived and authors now acknowledge the use of elements of social responsibility, of which some are drawn from the SRS (e.g. community, volunteerism, etc.) for use in more complex studies. However, recourse to Berkowitz and Lutterman's (1968) SRS is still implicit within many studies (e.g. Roberts 1993; 1996).

The 'green' consumer   More recent research has focused on the consumption attributes of the 'green consumer'. Stead et al. (1991, 833) have defined green consumers as '… those people who use their buying and investing powers to express their environmental concerns. They tend to buy products that are durable, have been produced with minimal damage to the environment and come in recyclable packaging'. Such sentiments are shared by Roberts (1993, 140) who states that the green consumer is one who '… purchases products and services perceived to have a positive (or less negative) influence on the environment'. These two definitions are, on the surface, very clear and do not provide any apparent cause for argument. However, there are two major concerns that can be raised following a close examination of the research into green consumption. In the first instance, the two quotations above clearly relate green buying to an intention to purchase 'green' items as a means by which to reduce environmental damage. However, green products may not be purchased primarily due to their environmental credentials but, in line with other rationales, may involve decisions that take into account moral or ethical dimensions (such as Fair Trade products). Accordingly, the assumption that 'green' buying is determined by solely environmental concerns is problematic. This difficulty is perhaps more acute with green consumption, but it is not restricted to this form of environmental behaviour. For example, buying recycled toilet paper or saving energy might be related as much to cost as environmental preservation. These difficulties imply that green consumption as a behavioural domain cuts across a range of other conceptual strands, driven by both environmental and other concerns. This leads to a second concern, that any specific scale of green purchasing behaviour will be inadequate for measuring the complexity of green buying.

However, from the studies reviewed, activities that could broadly constitute green consumption include:

- purchase of household goods (detergents/cleaners) with reduced phosphates or other pollutants to the natural environment;
- purchase of cosmetics (deodorants/sprays) with reduced levels of atmospherically harmful chemicals;
- organic products;
- source and packaging reduction;
- purchase of recyclable products.

As the reader will note, the last two categories also relate to waste management, illustrating points of cross-over between types of environmentally-responsible activity.

Conclusion

The classifications of environmental behaviour used above are contested; they can be critiqued from a number of perspectives and do not encompass every conceivable activity. Indeed, they do not incorporate the arguably critical field of sustainable travel. However, they provide a working basis for an examination of environmental action that is both manageable and reflects the current research and policy agendas in the field of study. Yet the focus of most research exploring the psychology of environmental action has been targeted less on debates surrounding *definitions* of behaviour but more on the types of factors that *influence* levels of commitment in these activities. It is to this topic that the chapter now turns.

As indicated in Chapter 4, three types of variables have been identified in social-psychological research which act to influence environmental behaviour:

- social and environmental values;
- situational variables;
- psychological variables.

These three sets of variables feed into the conceptual framework that has been developed to understand the alternative influences on environmental behaviour (Figure 4.3) and provide the basis for appreciating how such behaviour can be encouraged. This review of the literature is necessarily brief, but for more specific reviews for each type of behaviour, please see Barr and Gilg (2006), Barr et al. (2005) and Gilg and Barr (2006). This review therefore provides the basis for examining how these different variables relate to both different environmental practices and the extent to which these influences may vary when alternative lifestyle groups are identified.

Values

As described in Chapter 2, 'values' can be defined as the underlying guiding principles in people's lives (Schwartz 1992). Research into the relationship between 'values' and environmentally positive behaviour has been ongoing since the early 1970s and a close examination of the literature presents some significant problems for exploring the role of values and their influence on environmental behaviour. The main issue of concern centres on definitional problems. As studies by Maloney and Ward (1973) and Arbutnot (1977) demonstrated, the early development of measures to discern the relationship between values and behaviour often interchangeably utilised the concept of values with attitudes. In contrast to values, attitudes are generally considered to be specific and behaviourally-related views, whereas values are general principles. However, this confusion does necessitate the researcher to take a more critical perspective on the nature of values insofar as they relate to the environment and environmental behaviour.

Barr's (2003) framework for classifying values towards the environment (Chapter 2) is based on three worldviews. In the first worldview, social values provide the basis for an individual's orientation towards the value of the environment. This worldview utilises Schwartz's (1992) notion of 'universal' values to examine the role of social value orientations on environmental behaviour. The second worldview is conceptualised in terms of 'relational' values (Barr 2003). These values are specifically related to the environment and measure the hierarchical relationship that humans perceive they have with nature. Finally, the third worldview uses 'operational' environmental values (Barr 2003) to explore the extent to which individuals hold values consistent with ecocentric or technocentric approaches towards environmental management.

Social Value Dimensions

Seminal work by Schwartz and Blisky 1987) and Scwartz (1992) compiled research on social values from over 100 nations in an attempt to examine, in Schwartz's (1992, 1) own words, the potential for 'universals in social values'. In essence, this pertains to a search for a set of common social values across social and cultural boundaries. In brief, Schwartz has argued that there are two fundamental dimensions to social value orientations, the first being termed: 'self-enhancement–self-transcendence', representing what Cameron et al. (1998) have termed 'pro-self' and 'pro-social' values, respectively. The former represents an expression of self interest and the latter an expression of communal interest (Karp 1996), or what Schwartz (1977) also termed 'altruism'. The second dimension is termed 'openness to change–conservation' (note, this denotes wider social conservatism and not an ecological conservation ethic). The former represents a liberal/progressive position, the latter a status quo position. On the basis of these two overall dimensions, Schwartz outlines ten motivational value types, ranging from 'conformity' to 'self-direction', the scores on which can be used to identify individuals in terms of their overall position on the two value scales.

The first worldview that Barr (2003) explored related to how social values could be used as the basis for examining environmental behaviour. Stern (1992b) and Stern et al. (1995) have been the pioneers of this work in attempts to create new scales for measuring social values in relation to environment behaviour and found that individuals who scored highly on value types that emphasised 'pro-social' and 'openness' dimensions were more likely to undertake a range of environmental activities. Stern et al. (1995) argued that the nature of environmental action was likely to engage those individuals with pro-social or altruistic values more readily given the apparent lack of personal benefits over the wider and longer-term social advantages. Indeed, environmental behaviour, as a 'new social phenomenon' represented a challenge to the status quo, requiring a reconceptualisaton of personal priorities and lifestyle choice, supported in studies by Karp (2000) Corraliza and Berenguer (2000).

Relational Values

Moving to examine more specific environmental values, the second worldview identified by Barr (2003) pertains to the relational values individuals hold to nature. Relational values seek to examine how individuals view their role and relationship with nature and a number of authors (e.g. Dunlap and Van Liere 1978; Dunlap et al. 2000; Thompson and Barton 1994) have used the terms 'biospheric' and 'anthropocentric' as a means of creating a spectrum, which individual can be placed along. In the first instance, biospheric (or biocentric) values relate to values that emphasise the close relationship humans hold with nature. In its most extreme form, biocentrists do not make a distinction between the natural and human worlds arguing that humans have no more value than plants or animals. The anthropocentric perspective highlights a hierarchical and human-centred notion of the world, with human having greater value than either animals or plants. It is interesting to note that most of the major criticisms of recent sustainability policies have been based on an argument which highlights an anthropocentric viewpoint. This is evidently contested and there is by no means one type of anthropocentrism; most individuals do recognise the value of humans over nature, but do not see this in respect of 'dominion'.

This is a topic that has been investigated by a range of academics and most notably Riley Dunlap and Kent Van Liere, two American sociologists (Dunlap and Van Liere 1978; Dunlap et al. 2000). Dunlap and Van Liere's (1978) seminal piece of work was the construction of the *New Environmental Paradigm* (NEP) scale. In their 1978 article, Dunlap and Van Liere argued that American values towards nature were moving away from what they termed the Dominant Social Paradigm (DSP) towards the NEP. The DSP was characterised by over-consumption and a focus on material goods and human concerns. In contrast, the NEP represented a series of values, many of which display biospheric elements, that highlighted conservation and the importance of nature. Consisting of a 12-point Likert scale, the NEP used oppositional statements to gauge agreement. Examples of relational values included statements such as 'Humans were created to rule over nature' and 'Plants and animals do not exist primarily for human use'. Their research in Washington State found that there was higher than expected agreement with the biospheric elements of the NEP. The NEP has provided for a revolution in how environmental behaviour studies are undertaken. However, this relational value scale has been subject to criticism in a range of areas. Albrecht et al. (1982) and Kuhn and Jackson (1989) argued that the NEP scale was multidimensional and measured more than the single concept, which Dunlap and Van Liere (1978) had argued. However, more concerning was the role that the NEP played in predicting behavioural commitment towards a range of environmental actions. Vining and Ebreo's (1992) study of recycling in Illinois highlighted a potential deficiency in the NEP. They found that across their diverse sample of residents, there was generally high agreement with the NEP and that differentiating between 'recyclers' and 'non-recyclers' on the basis of the NEP was problematic. They noted that 'These results indicate that the differences between the recyclers and non-recyclers ... are a matter of agreement and not a contrast in fundamental values' (Vining and Ebreo 1992, 1603). Relational values, particularly in the form of he NEP, therefore provide a basis for examining environmental action within a wide context. Scott and Willits (1994) highlight a major difficultly when undertaking research that seeks to gauge individual values towards the environment, given the mass media's coverage of environmental problems. The overstating of environmental concern can cause survey results to be unreliable in some instances. However, it is likely that although scores on environmental value scales will be high, the differences between individuals will represent actual differences in values. By using values as indicators of general agreement with a set of biospheric–anthropocentric agreement and relating these to generic or composite measures of behaviour, the importance of relational values can be more critically examined.

Operational Values

Relational values plot individuals along a biospheric–anthropocentric continuum, but in many instances policy-makers interested in behaviour change wish to gauge the type of policies that hold the most sway with the population. Operational values can assists in this task, by demonstrating the underlying values that drive individuals' preferences for action to tackle environmental problems. O'Riordan (1976; 1985) has provided the basis for a conceptualisation of operational values along an ecocentric–technocentric spectrum, where ecocentrists represent a constituency who, driven by a biospheric set of relational values, argue that environmental policy should work within natural limits and constraints to development. In some instances, as Pearce (1993) has demonstrated (see Figure 2.6), a 'reduced growth' approach may be proposed by ecocentrists, as a way of reducing human-induced environmental impacts. Techocentrists argue, on the basis of anthropocentric values, that growth lies at the heart of a successful society. Accordingly, to ensure that growth can occur, a strategy based on passing from one generation to the next a suitable level of capital is either based on improving technological efficiency or exploiting more natural resources on the assumption that technological advances will eventually replace the need for these resources. A considerable amount of debate in this area inevitably centres on the nature of 'limits' (see Chapter 2), which forms an axis around which either pro-growth or anti-growth arguments turn.

The Role of Values: An Evaluation

Values, as guiding principles in peoples' lives, do provide the logical basis for behaviour. However, research has suggested that three problems underlie the study of values as useful determinants of environmental action. First, as Barr (2003) has outlined, values are complex and can be divided into a series of worldviews or domains (Arbutnot 1977) that seek to capture broad and often intangible concepts, such as the relationship humans have with nature. The rise of environmentalism and the status of sustainable development within the media and popular culture means that successfully measuring environmental values is becoming more problematic as people become versed in using the lexicon of sustainability (Scott and Willits 1994). This 'measurement' problem relates strongly to an issue that almost all studies have highlighted, which is the apparent discord between values and actions (Minton and Rose 1997). Two potential reactions to this problem can be considered. On the one hand, the social pressure to display apparently pro-environmental values may cloud the tangible differences that exist between values and actions. However, it may also be the case that the complexities highlighted by a number of researchers of the relationship between values and action are indeed present and that values are not significant in driving behaviour for many individuals. If this proposition is accepted, then a third problem can be identified. This pertains to the uncritical nature of a great deal of the research which has sought to examine the role of values in predicting behaviour. Barr et al. (2001) highlighted the potential for apparently specific environmental actions such as household waste management to occupy alternative behavioural domains (i.e. waste minimisation, re-use and recycling). Their research revealed that environmental values were significant in directly predicting waste minimisation behaviour, but not re-use or recycling activity. Barr et al. (2001) argued that this may have been due to the relatively new behavioural concept of waste minimisation, in contrast to the established and logistically more simplistic nature of recycling. Accordingly, individuals may act in accordance with their values when new behaviours emerge which are not governed by existing social norms.

Despite these reservations, values can provide a useful means by which to distinguish between individuals and their behavioural commitment. This can translate into the practical approaches that policy-makers take towards promoting environmental behaviour. For example, individuals holding pro-social, liberal, biospheric and ecoentric values could be targeted in very different ways to those who hold more conservative, egoistic, anthropocentric and technocentric values towards the environment. Having considered the role of values in predicting environmental action, we now turn to the second set of variables that influence environmental behaviour: situational variables (Figure 4.3).

Situational Variables

The term 'situational' is contested within the field of environmental psychology (Schultz et al. 1995), but is used within this text to refer to variables that provide context to an individual's setting (Figure 4.3). Even using this encompassing definition, there are a range of terms that have been used to examine these general set of variables. For example, the term 'structural' is often applied to situational variables that pertain to non-human factors that may underlie behaviour (for example, the amount of recycling provision provided by a local authority or the influence of a building's structure on saving energy). Another term used relates to 'background' variables (Stern and Oskamp 1987), which often pertain to 'human' elements (such as socio-demographic variables). Whichever of these terms is applied, researchers clearly distinguish these factors from the attitudes and perceptions of individuals, which we will examine in the next section.

Linking situational variables to environmental behaviour has been approached in a variety of ways by researchers. Early studies of energy and water saving (see Berk et al. 1980; McDougall et al. 1981; Ritchie et al. 1981; Stern and Oskamp 1987) highlighted the necessity of linking structural variables to observed energy use (such as recorded utility bills). These macro-level studies provided the basis for an overall and very accurate measure of the impact that weather and climate had on energy and water consumption. Indeed, there is little doubt that measuring observed behaviour provides the most accurate research material with which to ascertain current and changing levels of commitment to environmental action. However, this use of 'observed' data is problematic, both from a behavioural and situational perspective. First, the use of utility bills to gauge water or energy use (or for that matter, weighed recycled material) makes some key assumptions regarding the constancy of consumption within households, which the researcher is unlikely to be aware of. For example, utility bills as proxies for water use are useful if it is assumed that the household size is constant and that 'irregular' events can be controlled for. In both cases, researchers rarely have access to such data. Second, situational data relating to weather and climate can be of great utility in providing context for wider research findings, but the application of macro-level data without personal, attitudinal information is problematic when deriving policies for changing behaviour are sought. Indeed, a final difficulty for both situational and behavioural observed measures is that both rely on obtaining detailed and personal data from commercial companies. Under new data protection regulations in the UK and bearing in mind the sectional interests of commercial organisations, the use of such data is now almost impossible

Accordingly, researchers have tended towards the use of *reported* behaviour and situational variables as their main mode of data collection. Corral-Verdugo (1997) and Corral-Verdugo et al. (1999) have attempted to examine the potential differences between a respondent's reported behavioural commitment and the actual observed activity. In two studies of households in Mexico, they found that reported behaviour was consistently over-reported by their sample. However, although over-reporting was commonplace, the nature of this over-reporting was very consistent, leading the authors to conclude that researchers should use their measures as proportional, rather than absolute statements of behaviour. Accordingly, studies of the impact of situational variables on behaviour have moved onto a more common footing since the 1990s, with the dominant characteristics being studies as follows:

- structural variables (such as services provided and the built environment);
- socio-demographic characteristics (most notably age, gender, income, education, political affiliation and household size);
- knowledge and experience of environmental behaviour.

As noted above, studies of energy and water saving focussed on climatic variations as a key *structural variables*. However, the impact of dwelling size was also regarded as significant to the amount of resource used within the household. Stern and Oskamp (1987) placed these factors at the top of their hierarchical model of energy use behaviour, inferring that these refer to the size of dwelling unit and temperature conditions. Within the context of waste management behaviour, attitude measurement has been consistently aligned with the study of situational variables. Unlike energy and water saving, the key situational characteristic related to structural factors relates to recycling provision. As all readers of this text will appreciate, the range of recycling schemes operating at the household level is heavily dependent on location. Alternative local authority arrangements for recycling involve both kerbside recycling and civic amenity sites. Both Ball and Lawson (1990) and Barr et al. (2001) demonstrated the impact of location and recycling provision in the UK, with Barr et al. (2001) finding a linear relationship between increased levels of kerbside provision and recycling participation. In contrast, Guagnano et al. (1995) argued that for those without a kerbside facility, attitudes were a significant predictor of recycling activity.

Although there are nuanced differences between these studies, it is evident that a key structural characteristic influencing recycling activity is the provision of adequate facilities. However, although structural characteristics have been clearly demonstrated as significant predictors of recycling, energy and water saving behaviour, their impact on consumption activities has received less attention. Roberts (1996) has argued that the most important structural characteristic that can influence green forms of consumption relates to availability. The nature of green consumption as a minority market segment has meant that market development has been limited to areas where a demonstrable demand can be established. Accordingly, 'green' products have limited availability, subject to commercial demands.

The study of *socio-demographics characteristics* and environmental action is a long-standing venture and the meta-analysis provided by Hines et al. (1987) demonstrated the contested situation 20 years ago. The compartmentalised approach to the study of environmental behaviour has inevitably resulted in specific studies that have constructed socio-demographic profiles with little recourse to wider research on environmental action. However, researchers have continued to explore the role of variables such age, gender, educational background, income and political affiliation as either predictors or correlates with environmental action. Despite this situation, a review is well-overdue to consider the contemporary role of socio-demographic characteristics.

A final situational characteristic that researchers have examined relates to the *knowledge* that individuals receive and interpret towards the environment and the specific behaviour being studied. Schahn and Holzer (1990) have provided a useful perspective on the construction and reception of knowledge relating to the environment. They define two specific types of knowledge:

- *abstract:* knowledge that relates to the environment holistically. Such knowledge is often constructed as an 'awareness' of environmental problems;
- *concrete:* knowledge that relates specifically to changing a behaviour. This type of knowledge is operational, enabling the individual to take action to change their behaviour.

Schahn and Holzer (1990) argued that concrete knowledge was the most effective in changing behavioural practices, since abstract forms of knowledge related neither to the self nor a practical means of taking action. Examples of abstract knowledge proliferate the environmental psychology literature. In terms of energy conservation, McDougall et al. (1983) examined the role of 'eco-knowledge' and what Van Raaij and Verhallen (1983) termed knowledge of energy costs and consequences of these behaviours. Moore et al.'s (1994) study of water conservation also conceptualised knowledge relating to the water cycle and the overall use of water in certain household situations. These generic knowledge scales have been found to relate to levels of environmental awareness (e.g. Witherspoon and Martin 1992), but not to overall behavioural commitment. However, as Witherspoon and Martin (1992) note, there is a stronger relationship between specific behaviours and activities that are clearly achievable.

Related to environmental knowledge in the role of behavioural experience in shaping attitudes and practices towards new forms of environmental action. Daneshvary et al. (1998) have provided evidence that previous experience in one form of environmental behaviour can lead to a greater propensity to adopt other forms of environmental action. This is a significant consideration in developing a lifestyle profile for individuals and especially in targeting new policies.

Conclusion

Situational variables evidently play a key role in shaping environmental action. Structurally, issues of access, prevision and the built environment can influence the ability and opportunity of individuals to adopt environmental practices. However, the role of context, in the form of socio-demographic factors and knowledge, is evidently important. Social composition in itself does not necessarily provide a rationale for different levels of reported behaviour, but may represent the propensity of different social or lifestyle groups to adopt environmental actions at differing levels and rates. Evidently, attempting to compare across cultural and national boundaries as we have done above is problematic and may explain some of the contradictory findings. Such a review does remind the researcher that attempting to draw out crude generalisations from a wide variety of research contexts is to be avoided. However, as these studies have clearly demonstrated and as indicated in Figure 4.3, situational variables are only part of the issue and it is to psychological factors that we now turn our attention.

Psychological Variables

What can broadly be termed 'psychological variables' pertain to a range of attitudinal and perceptional influences that relate to a wide range of motivations and barriers to engage in environmental action. In most cases, it is not possible or desirable to uncouple factors on this basis, given that many influences can be both motivators and barriers, depending on the perceptions of the individual. For example, the 'convenience' of environmental action is viewed differently by individuals, depending on their perceived levels of service availability for a specific behaviour. The adoption of kerbside recycling can be seen as a great benefit by many, but a severe inconvenience to others, who have been used to placing all their refuse in one bin.

In a similar way to situational variables, psychological variables are heavily dependant on the behaviour in question. For example, research on energy saving behaviour has focused on the role of 'comfort' and the attitudes of individuals towards sacrificing some level of personal comfort to save energy (Seligman et al. 1979). However, there are some commonalities that run through the literature which will now be outlined.

Social Influence and Subjective Norms

The role of social influence in determining levels of environmental action has been highlighted from the 1970s, embedded as it was in the model promoted by Fishbein and Ajzen (1975). Fishbein and Ajzen initially considered that subjective norms were composed of an individual's awareness of a norm to act in a particular manner and the acceptance of that norm. Studies by a host of researchers have all indicated that the influence of what Oskamp et al. (1991) termed 'significant others' was important in generating higher levels of behavioural commitment. Some of the most prolific research in this area has been with regard to the influence of kerbside recycling on motivating others to participate (Gamba and Oskamp 1994; Tucker 1999). Kerbside recycling presents a means by which individuals can measure the environmental behaviour of others in their locality. The placement of recycling bins on the roadside on a regular basis presents residents with the ability to 'judge' the performance of their neighbours against themselves (Werner and Makela 1998). This opportunity that kerbside recycling presents has been utilised by policy-makers to create what Everett and Pierce (1991–92) term 'block leader' programmes. In the UK, this refers to forms of community-based recycling. In such schemes, community 'champions' act to motivate their community in attaining specific levels of recycling. The UK's most successful form of community recycling is the Community Composting Network, which is often used as a socially cohesive device within marginal or sparsely populated areas.

The role of social influence has also been examined with regard to other environmental actions. In terms of energy conservation, Costanzo (1986) has argued that 'social diffusion' offers a significant opportunity for individuals to display their habitual energy saving habits to others in the home and workplace. Indeed, Kantola et al. (1982) and Lam (1999) report that demonstrations of water-saving activity can impact significantly on others to take similar action. These studies evidently assume that displays of such activities are considered socially desirable by other individuals. Sadalla and Krull's (1995) study of environmental behaviour examined this notion of social desirability within an experimental setting. In a telephone survey, they provided respondents with two scenarios of a regular household routine. The only differences related to specific energy and water saving behaviours undertaken in one of the routines. Respondents were then asked a series of questions concerning the 'status' of individuals undertaking each of the two routines. With great consistency, the routine with energy and water saving behaviours represented individuals with lower social status.

Intrinsic Motivation

The role of 'motivation' in determining levels of environmental action has been highlighted in a series of papers by the American psychologist Raymond De Young. De Young (1986, 447) has argued that:

> People seem to derive considerable satisfaction from the very activities that others try so hard to encourage them to do [such that] ... conservation can also be perceived as contributing to one's sense of satisfaction

Drawing on work from nine previous research projects (see, for example, De Young 1985–86; 1988–89; 1990; De Young and Kaplan 1985–86), De Young (1996) characterises intrinsic motivation by five categories: frugality, participation, luxury, altruism and competence. It is worth briefly examining the fundamental elements of each of these categories which relate to those individuals intrinsically motivated to conserve.

In the first instance, De Young asserts that frugality is '... the prudent use of resources and the avoidance of waste' (De Young 1996, 371). It is a concept that can be incorporated into everyday life, regarding '... what items one buys, what activities one pursues and what one does with used or waste materials' (371). Second, participation supports notions of satisfaction from making a difference and being involved in worthwhile activities that have long-term benefits. Third, luxury, which might seem somewhat of an odd category to include with regard to conservation, applies to '... the satisfaction people feel in being members of a thriving affluent group' (378).

As such, this category is a recognition of personal yearnings for a comfortable life. Nonetheless, De Young argues that this category does have good correlations to other variables on the intrinsic motivation scale. Indeed, De Young argues that such a close relationship provides evidence against notions that '… conservation was the behavioral equivalent to freezing in the dark' (377). Fourth, altruism refers to the extent to which individuals are helpers, wishing to meet the expectations of others around them and alleviate genuine problems. Finally, competence refers to '… satisfaction derived from striving for behavioral competence' (379).

In the studies De Young outlines, all of the five satisfaction categories have relationships with conservation behaviour and are related to each other. It seems, as De Young has previously argued with regard to recycling (De Young 1986) that taking part in behaviours that are enjoyable and provide personal satisfaction can maintain what is overall repetitive and everyday action. Most crucially, De Young (1986) has argued that intrinsic motivation can be significantly more powerful than extrinsic forms of motivation. In his study of recycling behaviour, De Young (1986) demonstrated that those given extrinsic motives to participate in recycling (such as monetary rewards) only maintained their behaviour whilst the reward was available. Intrinsically motivated individuals therefore presented more consistent and long lasting behavioural commitments.

Environmental Threat and Perceived Severity

Implicit in many environmental campaigns are messages that highlight the potential impact of specific threats to both society and the individual. In contemporary British society, the impacts of climate change are often presented in apocalyptic terms, with predictions of sea-level rise, drought, increased storminess and energy crises. The use of negative campaigning to increase perceived levels of threat posed by negative environmental practices is used to 'shock' individuals into changing their behaviours in an attempt to avert disaster. Baldassare and Katz (1992, 604) have studied the impact of environmental threats, arguing that '… environmental threats overshadow youth, high education, high income and liberalism as predictors of overall and specific environmental practices'. Research by Steel (1996) and Segun et al. (1998) has also highlighted the role of perceived and specific threats to well-being in motivating environmental action amongst individuals.

Response Efficacy

Taking individual action to tackle environmental problems is often regarded, in colloquial terms, as being 'pointless'. A recent debate on BBC Radio 4's *Today* programme highlighted this issue in a discussion on the relative environmental benefits of 'eco-cars' being driven by the Prime Minister and the Leader of the Opposition. One environmental campaigner being interviewed remarked that the relative environmental benefit gained from an individual driving a slightly cleaner motor car was like '… farting in a hurricane, when compared to a plane load of tourists heading off to Tenerife on a 757'. This quotation encapsulates the perception that small actions to help the environment, both in and beyond the home, are perceived as lacking efficacy. This is bound up with a sense that unless a critical mass is involved in a specific activity, there is little point in participating. The impact of a positive sense of response efficacy has been widely researched in relation to a wide variety of environmental actions. However, given the long-term and somewhat intangible nature of energy saving impacts, it is not surprising that research has concentrated in this area (Becker et al. 1981; Verjallen and Van Raaij 1981; Midden and Ritsema 1983). Samuelsen and Biek (1991) highlight perhaps the most significant issue, relating to the temporal aspects of response efficacy and installing energy saving equipment such as solar panels, where even when individuals accepted the wider social and environmental rationales for participating, the perceived time for a 'payback' was regarded as too lengthy. This is reflected across the range of environmental actions, with Roberts (1996) demonstrating the impact of 'perceived consumer effectiveness' (PCE) on green consumption activity. This measure uses statements that question individuals about their power as consumers to reduce pollution and has consistently been found to influence consumption behaviour. Accordingly, in relation to both consumption and more habitual activities, raising levels of response efficacy is a major challenge for policy-makers seeking to encourage environmental action.

Self-Efficacy

Ajzen's (1991) adaption of the Theory of Reasoned Action into the Theory of Planned Behaviour (TPB) was predicated on the significance afforded to what Ajzen termed 'perceived behavioural control'. This 'locus of control' refers to the extent individuals feel empowered, or able, to undertake a specific behaviour. Such a definition appears to be narrow, but when placed into a wider context, refers to a range of factors that include perceptions of convenience, time, effort and other 'logistical' factors that may impact on the individual's perception of their ability to participate.

Many of these factors focus on 'convenience' and 'time' issues (Gamba and Oskamp 1994; Vining and Ebreo 1990), where perceptions of the specific requirements of the behaviour can determine participation. A useful example of this is provided by McKenzie-Mohr et al. (1995) in a study of composting behaviour. They differentiated between those who were willing and unwilling to compost garden and kitchen waste according to their perceptions of effort involved in dealing with compost. Behaviours such as composting can require considerably more effort than more habitual actions (e.g. switching off lights). Accordingly, the 'nuisance' factor becomes important in distinguishing between individuals who are willing to expend time and effort and those who will not build this into a daily routine. This has also been evident in relation to green consumption, where Schwepker and Cornwell (1991) and Sparks and Shepperd (1992) have demonstrated that those with high levels of self-efficacy were more likely to purchase a range of 'green' products, such as organic food.

Environmental Responsibility: Rights, Responsibilities and Trust

Consistent with the New Labour agenda of the 1990s and Etzioni's (1993) 'new communitarianism', the issues of environmental rights (Waks 1990) and responsibilities has been paid less attention than might have been expected. Studies by Nancarrow et al. (1996–97) and Lam (1999) have demonstrated the impact of these two distinct components of environmental citizenship (Selman 1996). Lam's (1999) study of water saving in Taiwan demonstrated the influence of environmental rights as a barrier to conservation behaviour. Those who regarded access to a constant and limitless supply of water as a right were the least likely to take measures to reduce water usage for the good of society. Indeed, one can relate arguments of rights and responsibilities back to Schwartz' (1992) notion of universal values, along the 'pro-self- and 'pro-social' continuum.

In relation to green forms of consumption, Berkowitz and Lutterman (1968) have used their social responsibility scale (SRS) as a proxy for green or responsible consumerism, but apart from this attempt to examine responsible consumption, there has been surprisingly little research using 'responsibility' as an independent variable in attempting to define the green consumer. The acceptance of personal responsibility also inevitably relates to the trust with which individuals treat environmental information. The claims of both government and environmental organisations are constantly debated in the media, leading to multiple discourses of environmental information. The wider decline in trust within British society must also be noted, with increasing scepticism of politicians and government. These wider social processes impact on the effectiveness of environmental information campaigns and blur the lines of communication between scientist, government and citizen.

Behaviourally-Specific Attitudes and Consumer Beliefs

There are further categories that are of significance when examining the psychological determinants of environmental behaviour, which mostly relate to green consumption and can be broadly defined as 'health and safety issues' and 'cost, convenience and range issues'. With regard to the first of these, Mackenzie (1990) predicted a growth in green consumer products and purchases during the 1990s which did not materialise. She argued that health and safety issues would be of such significance that there would be a massive shift towards organic products along with a rejection of the existing use of pesticides and fertilisers. Yet, as Roberts (1996) has noted, the green and organic boom was to a large extent a market flop. However, Mackenzie was correct in predicting the growth in concern over food quality and safety (if not the behavioural shift). Popular discourses regarding the safety of beef (emanating from the BSE crisis) as well as the relatively new controversy over GMO's have been raised in the public consciousness.

The second issue is of more relevance to a wider set of green consumer behaviours. Studies by Shrum et al. (1995) and Mainieri et al. (1997) have shown how the importance of price, quality, product loyalty and the interest and care in purchases can impact on the behaviour of consumers. In the study by Shrum et al. (1995) a lifestyle survey was used to examine attitudes towards buying biodegradable packaging and switching brand to a more environmentally safe product. They found that those who were interested in new products and took care when shopping were more likely to both buy biodegradable products and switch brands. No effect for brand loyalty was seen. Mainieri et al.'s (1997) study of green buying provided some evidence for the significance of price and quality influencing choice, but again the data presentation does not allow sufficient data disaggregation. Nonetheless, an experimental study by Schuhwerk and Lefkoff-Hagius (1995) does provide evidence that green rationales for purchasing products can be more important than financial ones, especially for those with low involvement in the environment. In their study, they provided individuals with information on two brands of 'green' washing powder, the only difference being between the promotional basis. One was based on green motivations to purchase the powder, the other on financial benefits. The authors found that those who were less involved in the environment generally were more susceptible to green arguments as opposed to financial ones. It does appear that there may be some variability concerning the efficacy of price consciousness in modifying behaviour, although Verhallen and Van Raaij (1981) have shown the importance of price concerns with regard to energy conservation.

Conclusion

The psychological variables examined in this section provide a brief overview of the many factors that researchers have linked to environmental action. Crucially, they are drawn from a disparate set of research agendas, encompassing a wide range of behaviours. However, they contain many commonalities which can be used to address the issue of behavioural change. They combine both external factors (such as subjective norms) with internal factors (intrinsic motivations) to formulate a set of psychological variables that can be evaluated within the context of both situational variables and social and environmental values. We now return, therefore to the overarching conceptualisation of environmental behaviour that forms the basis of the empirical research reported in Chapters 6 to 8.

Conceptualising Environmental Behaviour: Intentions and Actions

Figure 4.3 presented a basic framework for conceptualising environmental action. As described in Chapter 4, this framework is designed to act as a conceptual device for organising variables and does not present a 'model' of behaviour. As the review above has demonstrated, the nature of environmental action implies that multiple forms of this framework are needed for specific activities. This complexity in environmental action, both in terms of its definition and potential influences highlights the need for policy-makers to engage in the debates surrounding behaviour change. Following a review of both the structure of environmental behaviour and the myriad of influences upon it, there are a number of key issues that emerge from this chapter. First, behavioural commitment is conceptualised at present along highly compartmentalised lines, with both policies and academic research utilising strictly vertical forms of understanding. One of the difficulties of conducting a wide ranging literature review is the alternative discourses used within the specific studies which make comparison across behaviour problematic. Yet research does demonstrate that the type of factors which influence behaviours are similar, leading to the possibility that environmental action is framed in terms of alternative conceptualisations of everyday practice.

A second key issue to emerge from this chapter and related to the former point, is that alternative environmental practices, however they are conceptualised, are influenced by a range of differing factors:

- *values*, as underlying guiding principles in peoples' lives, appear to have less direct impact on environmental practices, yet may form the basis for indirect effects on attitudes. In contrast, for some new forms of environmental action, values may be a very important motivating factor in the first instance;
- *situational variables* are heavily contextual and behaviourally dependent, with specific levels of facilities and physical infrastructure having a major impact on behaviours. However, studies are equivocal, for example regarding the role of socio-demographic factors. This may partly be due to the broad behavioural conceptualisations used, which do not expose the underlying social or lifestyle dynamics of environmental practices;
- *psychological variables* are varied and mainly contextual, but do provide the ability to chart the alternative role of perceived barriers and motivators for specific behaviours.

A third and final point to emerge from this review relates to the overall approach that has been adopted by researchers in this area of study. Considerable research effort has been ploughed into examining the determinants on specific types of activity. Yet the significant (albeit confusing) demographic differences highlighted above may point to an alternative approach to examining environmental behaviour. The discipline of marketing has long utilised the technique of segmentation as a way of identifying the potential role of lifestyle groups. What much of the research reported in this chapter alludes to is the potential presence of different environmental lifestyle types, which may form the basis for a more effective study of environmental action. The next chapter takes these ideas and demonstrates how they can be tested and utilised in a primary research setting through the use of both quantitative and qualitative research techniques.

PART 3
Approaches

Chapter 6

Framing Environmental Practice

Introduction

The preceding chapters have demonstrated a gap in both the conceptual and policy-related literature that has emerged, and encompasses three inter-related problems in the way that environmental action has been both framed and strategies for its promotion employed. The first of these problems relates to the framing of environmental practices in and around the home. As we have seen in Chapter 5, environmental practices relate to those everyday actions of consumption and habit that can be undertaken in and around the home on a regular basis. Traditionally, researchers have conceptualised environmental practices within rigid disciplinary boundaries (such as 'energy studies' or 'recycling research'). This scholarship has detailed the specific nature of a given set of environmental practices, each related to an environmental problem. Indeed, from a practical perspective, policy has compartmentalised environmental practices into a series of problem-focused strategies. Exemplified by the *Are You Doing Your Bit?* campaign (DETR 2000), environmental practices have been conceptualised by policy-makers as distinct behaviours that are, in the consciousness of individuals, framed in terms of the external environmental issues that they represent. However, neither academics nor policy-makers have sought to realise the potential benefits of studying environmental practices within the wider conceptual or practical context in which everyday consumption and habits are practiced. For example, is it possible to conceptually relate activities such as 'green consumption' to energy and water saving practices which also involve consumptive practices? Indeed, is it probable that environmental practice is a reflection of everyday habitual and consumptive practices?

The second 'problem' has related to the assumption within the scholarly and policy community, that policies and programmes aimed at changing behaviour can be targeted at the population as a whole. Chapter 3 has detailed the range of sustainable development policies that have sought to employ what we have termed a 'top down' approach to behaviour change, with decisions on the types of messages to be presented to the public formed at a high level. The key assumption within national environmental campaigns such as 'Helping the Earth Begins At Home' and 'Going for Green' has been that messages of behaviour change will have a widespread impact. Yet evaluations of these programmes (Hinchliffe 1996; Collins 2002) have demonstrated the widespread *non*-adoption of behaviour change measures by the general public, who have often rejected the messages present in the programme as being irrelevant to their own circumstances. In other words, the promotion of 'green' messages has been received as an issue for 'someone else'. Accordingly, environmental practices are framed as both irrelevant and socially *un*desirable. This difficultly represents a wider challenge for social researchers and policy-makers seeking to encourage alternative forms of behaviour. Yet an element of the challenge involved pertains to the broad focus of behaviour change campaigns, which do not seek to discern differences between alternative target groups in society. Marketing approaches to behaviour change would argue that targeting policies might yield better results. In other words, is it possible that alternative environmental lifestyle groups can be identified and that differing policies can be targeted at these segments?

Finally, the third problem relates to the 'logic' of behaviour change. As Chapter 4 demonstrated, policy-makers have employed a 'linear' model of behaviour change, which has assumed that there is an implicit link between awareness, information, a decision to act and, finally, an action (A-I-D-A). This approach to behaviour change has been widely critiqued (see Chapter 4), but alternative perspectives from the geographical literature have been found to offer little more than a sophisticated culturally-informed critique of these models, offering less to the study and promotion of practical behaviour change policies. However, alternative approaches, grounded in social psychology, have been developed that can offer a new and holistic perspective on environmental action. In particular through the use of conceptual frameworks based on Fishbein and Ajzen's (1975) and Ajzen's (1991) Theory of Reasoned Action/Planned Behaviour, the central issue of intentions and actions can be examined. As Figure 4.3 demonstrates, environmental action can be conceptualised in terms of this discord between an intention to act and behaviour, with three key sets of variables acting to influence this relationship (social and environmental values, situational and psychological variables). This development of a framework for environmental action represents a significant step forward from the linear-models approach, providing researchers and policy-makers with a clear perspective on how environmental action is influenced. Building on the three previous 'problems' identified above, this approach can be used to examine the influences on both alternative environmental practices and different lifestyle segments, providing the means by which to target behaviour change policies at specific groups in the population.

This chapter outlines the basis for operationalising the three approaches detailed above for tackling these 'problems'. This is based on a major UK Economic and Social Research Council (ESRC) funded research project which ran from 2001 to 2003 (Barr et al. 2003)[1] and DEFRA funded research that was undertaken between 2005 and 2006.[2] The broad aim and objectives of these research projects will be outlined in the next section, from which the chapter proceeds to detail the methodology employed in the two related projects. The

---

1   The ESRC research (Grant No. R000239417) was led by Andrew Gilg and Dr Nicholas Ford from the University of Exeter.

2   The DEFRA research (Grant No. EPES050614C) was led by Dr Stewart Barr and Prof. Gareth Shaw from the University of Exeter.

next section of the chapter examines the basis for examining the first problem identified above, namely how environmental practices are related to each other and how, in turn, these relate to everyday practices in and around the home. Within this section, details of reported environmental action are provided for a range of forty environmental behaviours, demonstrating the nature of contemporary environmental practice in Britain. Finally, the chapter finishes by exploring the potential for using these new forms of environmental practice as the basis for formulating and promoting behaviour change policies.

Environmental Action in and around the Home

The motivation to explore the research problems outlined in the introduction to this chapter emanated from a survey of household waste management behaviour in Exeter during 1999 (Barr 2002; Barr et al. 2001). This research had demonstrated that sustainable waste management behaviour was clearly divided into three distinct elements: waste minimisation, re-use and recycling. Each of these three types of waste management practice had very different antecedents, with recycling behaviour influenced most by the provision of kerbside recycling, whilst waste minimisation activities were predicted by a range of social and environmental values and psychological variables. This research highlighted the importance of classifying behaviour according to the types of practices involved and led the research team to question the basis of studying environmental action on a sectoral basis. Accordingly, the research team applied to the Economic and Social Research Council (ESRC) of the United Kingdom for a £119,000 grant to undertake a two year study of 'Environmental Action in and Around the Home' (Barr et al. 2003). In a bid to extend this research and apply the findings more practically, the research team also tendered successfully for a £21,000 DEFRA research grant in 2005 (Barr et al. 2006; Barr et al. 2007a; 2007b). This latter project involved further analysis of the ESRC data and further primary qualitative research which will be detailed in Chapter 9.

Aim and Objectives of the Research

Although the projects were separate, they both sought to answer related objectives. Accordingly, the conflated aim of these research projects was to use the conceptual framework of environmental behaviour (Figure 4.3) in order to examine the potential for using lifestyle groups for changing behaviour through the identification of barriers and motivations for environmental action. The three empirical objectives of the research were:

- to identify the types of *environmental practices* people undertake in and around the home;
- to segment individuals according to their behavioural commitment into *lifestyle groups*;

- to examine the *barriers and motivations* for specific environmental practices and lifestyle groups.

These objectives directly addressed the three 'problems' identified at the beginning of the chapter and represented distinct empirical dimensions to the research, employing both quantitative and qualitative methods.

Methodological Approach

The first research project, funded by ESRC, was a major two year research grant involving a large quantitative study of environmental actions and attitudes. This research involved examining four key environmental actions: energy saving, water conservation, recycling and green consumption. The research team had debated the merits of adding travel behaviour into the research grant application, but eventually decided against its inclusion given the multidimensional and contested nature of transport research, which is a field in its own right. Although the team would have preferred its addition to the study, it was felt that this may compromise both the likelihood of funding and also the ability of the project to deliver a succinct and reliable survey.

The first six months of the research identified key literature in the fields of energy, water, waste and green consumption and used this work to modify the conceptual framework of environmental action (Figure 4.3). In order to empirically address the three research objectives, the research team followed existing practice within the social-psychological literature and based the design on a quantitative study instrument. This would ask questions relating to all of the variables in the conceptual framework, with behaviour being based on reported environmental action.

Study Area

The study location was Devon, a county in southwest England. Devon has a population of 704,493 (National Statistics 2007) and is predominantly a rural county, with a number of distinct geographical zones. Located on the South West peninsular (see Figure 6.1), the county in bounded by the Bristol Channel to the north and English Channel to the south. South Devon comprises a number of seaside resorts, such as Torbay and Sidmouth, with the northern coast being more remote. The western part of Devon is dominated by Dartmoor National Park, a high moorland environment. The eastern part of the county is comprised of rolling hills and rich agricultural land. The only major conurbation in the county is Plymouth (population 240,720; National Statistics 2007), a naval port dominated by its dockyard and attendant industries. The administrative centre of the county is the cathedral city of Exeter (population 111,000), a sprawling suburban city, economically driven by its university and a growing service sector, with significant inward investment being witnessed recently aided by the fastest expanding international airport in the UK. The rest of Devon is comprised of small rural settlements and market towns with a wide catchment area. Overall, Devon represents a distinct geographical location for research and is, by definition, not representative of the UK. However, its contrasts provide a useful set of study tools.

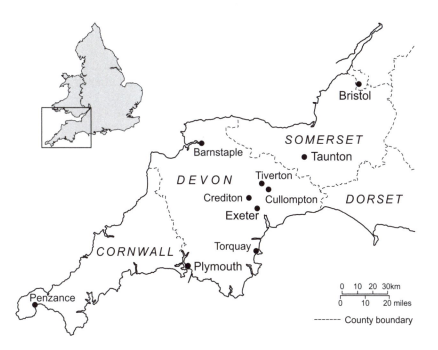

Figure 6.1   Study locations in Devon, South West England

The research team decided to base their research design and sampling framework on an urban-rural continuum. The hypothesis lying behind this design was that the influence of the built environment may have an influence on environmental practices, with those living in high density terraces having different barriers and opportunities for undertaking environmental action from those living in detached houses with gardens. Evidently, the urban-rural continuum is a crude generalisation, but it did offer a logic and practical means by which to design the sampling framework. Accordingly, the study locations were Plymouth, Exeter, Barnstaple and mid-Devon. Barnstaple is a market town in the north of the county, servicing a wide area of the north and comprising mainly sub-urban housing. Mid-Devon is a rural district to the north of Exeter, with small market towns (Cullompton, Crediton and Tiverton) servicing a mainly agricultural or commuting-based community.

Sampling Framework

Within each of these four study areas, it was decided that 400 households would be sampled at random, each being offered the opportunity to complete a questionnaire measuring the constructs in Figure 4.3 (see the following section). The sampling strategy employed followed Gray's (1971) recommendations concerning the selection of samples from official data sources. In this instance, the electoral register was used as the sampling frame, since this document had, in 2002, all names and addresses in each district within it. Using a systematic random sampling method, 400 addresses in each study area were selected. It is interesting to note that since 2004, access to the electoral register on this basis has been restricted, with householders now having the option not to appear on the main register, their details being kept privately in accordance with the Data Protection Act (1999).

Given the policy-related nature of the research, a research team member contacted all of the local authorities in whose locations the questionnaire would be delivered. Several meetings were held with officials from each authority, at which views were gauged concerning the types of questions that could be included, in addition to the essential academic requirements of the survey. These constructive meetings ensured that the survey had endorsement from all of the local authority's involved providing an opportunity to provide them with key data and enabling the researcher's to nuance questions that related to specific localities.

During the summer of 2002, an independent research company (Quests) hand-delivered each of the 1600 surveys to the pre-selected addresses. When no one was at home or a resident refused to complete the survey, the next property was sampled, until a response was gained. The normal list of addresses was then resumed. The questionnaire was then collected a few days later. This 'contact and collect' approach is very labour intensive and therefore costly: the sampling of 1600 households in Devon took over four months to complete. However, response rates were over 80 per cent in some locations and this benefit far outweighs the costs involved.

Study Instruments

The questionnaire that was hand-delivered to the 1,600 addressees in Devon was drafted over four to five months. The conceptual framework (Figure 4.3) determined that a wide range of variables had to be measured in order to capture the many concepts included in the framework. The major issues related to the behavioural questions in the survey. As Corral-Verdugo (1997) and Corral-Verdugo et al. (1999) have outlined, the use of reported behaviour as a measure of actual behavioural commitment is problematic, although relatively, the relationships between observed and reported action is fairly consistent. The ESRC research necessitated the measurement of activities such as energy saving, water conservation and green consumption. Although theoretically it might be possible to measure such behaviour by monitoring energy and utility bills, this does not reveal the behavioural commitment shown by individual members of a household nor reflect the actual activities they area engaged in. Accordingly, reported behaviour was selected as the measure of environmental action, with all of the qualifications that previous work has pointed to.

However, both the measurement of environmental behaviour and the types of activities to be included in such measures differ between previous research projects. Some projects have used simple 'yes/no' responses to behavioural questions, whilst others have utilised semantic differential scales to assess behavioural commitment. Indeed, given that behaviours such as 'recycling glass' and 'turning off lights' would be undertaken with alternative frequencies, the use of absolute measures of behavioural commitment (such as 'every day', 'every week', etc.) is also problematic. Accordingly, the research decided that temporally abstract measures of behavioural commitment would be utilised using a five-point frequency Likert scale: 'always', 'usually', sometimes', 'rarely', 'never'. This scale had the advantage of being both relatively nuanced and could also relate to a series of behavioural commitments that were not tied to absolute timings.

Measuring Behaviour

Deciding on the types of activities that would comprise measures of environmental behaviour was also challenging. Previous studies had utilised a wide range of activities, many of which were common across cultural and national examples, but others which were specific to a particular location or setting (see Chapter 5). Indeed, there was the wider issue that related to the potential conflict that researchers face between conceptual and practical measures of environmental action. To overcome this difficultly, the researchers decided to undertake a meta-analysis of 'recommended actions' from promotional materials provided to residents in the four study locations. This involved a content analysis of promotional literature and structured programmes, such as recycling collections. As seen in Chapter 3, local authorities have become the conduit for information on environmental behaviour, with programmes related to cutting energy usage, saving water and promoting green consumption and recycling. These responsibilities have their origin in the local authority role envisaged under Local Agenda 21 and the new duties that authorities have in promoting sustainable communities and Community Strategies.

Table 6.1 provides a summary of the activities that were selected. As can be seen, these were initially classified into the four key areas of interest (energy, water, waste and green consumption). In terms of energy saving, the ten behaviours identified pertained mostly to activities within the home, such as turning off lights and putting on more clothes rather than turning up the thermostat. However, two behaviours were consistently mentioned across promotional materials which were related to longer-term decision-making: looking for energy-saving appliances and purchasing energy-saving light bulbs. Water conservation also involves the use of both regular activities, such as turning off taps when cleaning teeth or washing dishes, alongside more long-term decisions, such as the purchase of a water butt. Indeed, water conservation also contained some longer-term home-based decisions, such as flushing the toilet less and using less water when showering. Green consumption activity is perhaps the most varied of the four behaviours, with a wide range of activities listed. This reflects the range of behaviours recommended by local authorities (varying as they do from the purchase of fairly traded goods to the avoidance of aerosols in spray cans). However, this also reflects the confused status of green consumption as a concept. As noted in Chapter 5, studies have interpreted green consumption within a wider social context, emphasising 'socially responsible consumption' (Roberts 1996). As Seyfang (2003) has noted, green consumption can incorporate notions of environmental and ethical consumptive practices, with motivations for these practices relating to a myriad of influences (Winter 2003). For example, although the purchase of locally produced food may imply a green motive (reducing the number of food miles involved), the production practices used may be less 'green' than for produce brought from further afield. Indeed, notions of 'green' also conflict with ethical dimensions; is it more sustainable to purchase bananas from Israel or fairly traded bananas from further afield (such as Columbia)? Indeed, these ethical and environmental motivations also conflict with economic dimensions of sustainable development. For example, Winter (2003) has argued that the promotion of local food, whilst ostensibly representing environmental benefits, can also reflect what he terms 'defensive localism'. This refers to a local economic imperative, which can take precedence over environmental or ethical concerns. Accordingly, 'green' consumption is highly contested and this study, whilst recognising these limitations, utilised the notions of green consumption used by local authorities in Devon. Finally, the research focused on waste management behaviours, which were mostly focused around recycling behaviours, but also included the donation of clothing and household items to charity. This sought to recognise the increasing importance of the charity sector in handling large amounts of materials that householders have finished using, which can be re-used without any mechanical processing.

Table 6.1   Reported behaviours measured in the survey

Energy-saving behaviours	Reduce heating in unoccupied rooms
	Reduce hot water temperature
	Wait for a full load before using washing machine
	Put on more clothing before turning up heating
	Switch off lights in unoccupied rooms
	Keep heating low to save energy
	Use high efficiency light bulbs
	Use high efficiency appliances
	Energy efficient boiler or 'Economy 7'*†
	Insulation (windows, doors and/or loft)*
Water conservation behaviours	Use a shower or fill the bath half full
	Reduce the times the toilet is flushed
	Reduce the number of baths and/or showers taken
	Turn off the tap when cleaning teeth

Table 6.1 cont'd

Water conservation	Turn off the tap when washing dishes
behaviours (cont'd)	Turn off the water when 'soaping up' in the bath or shower
	Use a sprinkler sparingly, if at all, in the garden
	Buying plants that require less water
	Use of a toilet flow controller or 'hippo'
	Use of a water butt
Green consumption	Use environmentally-friendly detergents
behaviours	Avoid aerosol products
	Purchase items with as little packaging as possible
	Use own bag when shopping, rather than a plastic one
	Buy recycled toilet paper
	Buy recycled writing paper
	Buy organic produce
	Buy Fair Trade products
	Use a local food store
	Buy local produce
Waste management	Take clothes to charity shops
behaviours	Donate old household items to charity
	Re-use glass bottles and jars
	Re-use paper
	Recycle glass
	Recycle newspaper
	Recycle cans
	Recycle plastic bottles
	Compost kitchen waste
	Compost garden waste

Notes

Behaviours were measured on a frequency scale of 1 (never) to 5 (always), except where *
indicates that these were yes/no answers.

† Economy 7 is a form of heating which uses electricity at off-peak times and stores it
to use during peak hours.

Behavioural Intention

The questionnaire also needed to contain items that could accurately distinguish a notion of reported behaviour from behavioural intuition (see Figure 4.3). This presented specific difficulties, not least because experimental psychologists have used behavioural intention as a reported measure and behaviour as actual, observed behavioural commitment. To overcome this problem, the research team decided to conceptualise behavioural intention as a 'willingness' to undertake a particular activity. However, although this provided a reasonable solution to this challenge, behavioural intention is also normally directly measured against a relevant behaviour. Given the constraints on length, it was not possible to have an additional forty items relating to a willingness to undertake each reported behaviour. Accordingly, a series of relatively abstract 'willingness' statements were used (see Table 6.2) that attempted to capture the essential elements identified in the differences between the behaviours.

Table 6.2   Behavioural intention items measured in the survey

Energy-saving behaviours	Invest in energy-saving devices
	Make changes in everyday life to save energy
	Purchase products that save energy
Water conservation behaviours	Make changes in everyday life to save water
	Make more lasting changes to save water
	Reduce water use in the garden
Green consumption behaviours	Look for environmentally-friendly products
	Change shopping habits to help the environment
	Make an effort to buy produce locally
Waste management behaviours	Recycle household waste
	Compost household waste
	Donate unwanted items to charity, rather than disposing of them

Note

Behaviours were measured on a 'willingness' scale from 1 (very unwilling to undertake) to 5 (very willing to undertake).

Social and Environmental Values

As underlying principles in people's lives, values form an important measure for research into environmental behaviour. This research defined three types of values that were of significance: social values and relational and operational environmental values. In the case of social values, the research team utilised the two dimensions proposed by Schwartz (1992). These relate to the continua of 'pro-self –pro-social' and 'open to change–conservative'. Stern et al. (1995) have argued that those more likely to help the environment were pro-social and open to change. In order to test this hypothesis, Schwartz's (1992) inventory of values was utilised to measure agreement with these concepts. These are listed in Table 6.3. As can be seen, these somewhat abstract concepts do not lend themselves easily to a succinct evaluation by individuals and so respondents were asked to provide their 'gut reaction' to the statements, ranking each one in terms of its importance to their lives.

Table 6.3   Social values measured in the survey

Unity	Social justice
Authority	Wealth
Exciting life	Curiosity
Honouring parents	Social order
Equality	Influence
Social power	Helpful
Varied life	Enjoying life
Obedience	Loyalty

Note

Individuals were asked how important each value was to their own life, rating each from 1 (very unimportant) to 5 (very important).

Relational and operational values were measured using an adapted form of Dunlap and Van Liere's (1978) New Environmental Paradigm scale (NEP) (Steel, 1996), as shown in Table 6.4. Respondents were asked to tick their agreement to each statement on a five-point Likert agreement scale. This amended NEP combines both biocentric-anthropocentric and ecocentric-technocentric items to provide a distinction between relational and operational values.

Situational Variables

Figure 4.3 demonstrates that both behavioural intention and behaviour can be influenced by situational variables. These were classified according to specific structural contexts, socio-demographic information and knowledge. In terms of structural variables, each local authority had different arrangements for the provision of recycling services, which necessitated alternative questions for each study area. For example in the city of Exeter, recyclable waste is placed into a green wheeled bin and is collected fortnightly, with non-recyclable waste being collected

Table 6.4   Environmental values measured in the survey

New Environmental Paradigm item 1	The balance of nature is very delicate and easily upset by human activity
New Environmental Paradigm item 2	The earth is like a spaceship with limited room and resources
New Environmental Paradigm item 3	Plants and animals do not exist primarily for human use
New Environmental Paradigm item 4	Modifying the environment for human use seldom causes serious problems
New Environmental Paradigm item 5	There are no limits to growth for countries like the UK
New Environmental Paradigm item 6	Humankind was created to rule over the rest of nature
Ecocentric value 1	One of the most important reasons for conservation is to preserve wild areas
Ecocentric value 2	Exploitation of natural resources should be stopped in order to protect the environment
Technocentric value 1	Technological advances will solve many environmental problems
Technocentric value 1	Developments in science mean that we will be able to maintain our standard of living without having to conserve

Note

Agreement for each statement was measured on a scale from 1 (strongly disagree) to 5 (strongly agree).

on alternative fortnights. Garden waste is collected separately. In contrast, North Devon (including Barnstaple) has weekly collections, but recyclables have to be placed into different containers according to their composition (e.g. glass, card, etc.). Such differences necessitated the use of slightly different survey instruments that would ascertain the level of provision (i.e. kerbside/civic amenity site) that households had access to.

Socio-demographic factors were measured using a range of questions that asked the respondent for information relating to their age, gender, political affiliation, education, income (in categories), membership of organisations, size of household, type of property, access to a garden and car access. These standard measures were used to compare across to other studies that have found socio-demographic factors to be significant in predicting environmental action. All were measured using tick box items, apart from age, where an absolute figure was requested.

In terms of knowledge, using Schahn and Holzer's (1990) recommendation, measures of both abstract and concrete environmental knowledge were used. These utilised Christie and Jarvis's (1992) environmental 'quiz', used in the British Social Attitudes surveys. The quiz contains ten 'true or false' items (Table 6.5) relating to both global environmental issues and local concerns. Many of the items relate to common misunderstandings of environmental problems, such as the causes of global warming. This 'abstract' form of knowledge was complemented in the questionnaire by the use of 'policy knowledge' items, where individuals were asked whether they had heard of a range of national and local environmental initiatives and, in addition, which sources they used to gather information about the environment (Table 6.6). These were drawn from local authority and national sources and individuals were provided with the opportunity (at the back of the questionnaire) to gain further information about specific environmental initiatives through the use of telephone numbers or websites.

Table 6.5   Environmental knowledge quiz in the survey

Statement	Answer
Nine-tenths of the hottest years have been since 1983	TRUE
Warming in the atmosphere is mainly due to increases in carbon dioxide produced from burning fossil fuels	TRUE
The Greenhouse Effect is caused by a hole in the Earth's atmosphere	FALSE
Road transport produces few harmful emissions into the atmosphere	FALSE
Under one-tenth of household waste is recycled	TRUE
There is a crisis in finding landfill space in Devon	TRUE
Two thirds of electricity in the UK is generated from coal and gas	TRUE
The UK uses some of the smallest amounts of renewable energy in Europe	TRUE
A quarter of household water is used flushing the toilet	TRUE
Personal washing uses very little household water	FALSE
The average household throws away 10kg of packaging a week	TRUE
Most food we buy in shops is organically produced	FALSE

Note

Respondents were asked to state whether each statement was true or false.

Psychological variables

Figure 4.3 provides the basis for the psychological variables measured in the questionnaire. Although there is not space here to justify each variable included, some of they key items can be highlighted. In the first instance, the role of norms

Table 6.6   Sources of information about environmental issues in the survey

BBC TV
ITV Carlton Westcountry
City Council leaflets
Exeter Citizen (City Council magazine)
Gemini FM (independent local radio station)
Express and Echo (local newspaper)
National press
Magazines
Radio Devon (BBC local radio station)
The Leader (free local newspaper)

Note

Respondents were asked to indicate how many of the above they used to obtain knowledge regarding environmental action.

and social desirability was measured using standard subjective norm items, asking whether respondents were aware of others around them helping the environment and if others engaging in environmental action would influence them to do so. However, in order to examine the notion of social desirability highlighted by Sadalla and Krull (1995), a measure of the 'desirability' of undertaking environmental action was developed. This caused significant problems for the research team and demonstrates that many of the underlying attitudes towards environmental action cannot be simply measured in a survey. The item finally arrived at read 'People who help the environment are a bit eccentric'. Although this is a fairly weak statement, it attempted to gauge whether individuals felt helping the environment was an aspirational goal.

A further factor of interest in terms of psychological variables related to self-efficacy and the 'locus of control' that individuals felt in undertaking environmental action (Ajzen 1991). This was operationalised through a series of measures in the questionnaire, relating to perceptions of time to undertake environmental behaviour, the convenience of environmental action and the physical realities of undertaking environmental practices (such as the room to store recyclable materials). The survey also measured response efficacy, examining a series of beliefs about the outcomes of environmental behaviour. Four specific items were created which provided outcome scenarios related to each behaviour of interest. For example, one item read 'energy saving in the home helps reduce global warming'. These items aimed to establish a potential link between behaviour and outcomes. In addition, statements also referred directly to response efficacy: 'Since one person cannot have any effect on natural resource problems, it doesn't make any difference what I do'.

Finally, intrinsic motivation was measured in order to test De Young's (1996) thesis that the satisfaction gained from helping the environment could overcome negative attitudes or structural barriers to this behaviour. This concept was measured according to a series of statements, such as: 'It makes me feel good when I do something to help the environment, such as reusing or recycling things'. To counter this type of statement, measures of extrinsic motivation were also developed, asking respondents about their attitudes towards receiving financial incentives/penalties for environmental action.

A large number of items were included to measure all of the constructs in the lower section of Figure 4.3. These were all measured on five-point Likert agreement scales. For each concept measured, a check statement was inserted, expressing the opposite attitudinal direction to ensure that internal consistency of responses could be assessed. In addition to the general attitude measures indicated in Figure 4.3, more specific attitudes pertaining to green forms of consumption were also included, such as the importance of health, safety and environmental concerns and, crucially, the role of price in determining green purchases.

The Final Questionnaire and Survey

The finalised questionnaires were 14 pages in length, using predominantly tick box items and utilised the major design principles of Dillman (1978). Questions were carefully ordered to elicit the most reliable responses as possible. For example, reported behaviour questions were placed at the front of the questionnaire, with the next section dealing with socio-demographics. The more challenging items relating to values then followed, with the psychological items being placed last. It was hoped that asking for reported behaviour at the start of the survey would elicit the most reliable response, rather than placing it after a series of attitudinal questions. The title of the survey was 'Environmental Issues in … [specific location]'. Although the use of the word 'environmental' did hint at the nature of the survey, it was felt that to use any other term would be misleading. The front cover used the university's logo, alongside the local authority's logo and any other branding that they had been using. The introduction to the survey emphasised the independent nature of the research, the confidentiality of data and where further information could be gained. In addition to this information, a cover letter was provided that explained the rationale for the research, signed by a member of the research team.

Both the covering letter and the questionnaire itself requested that one person only completed the survey. This presented the research team with a further challenge. Given that there was little control over who would actually complete the survey, the results would be representative of that individual, rather than the household. Yet many of the activities examined in the survey were essentially household behaviours, such as recycling. Accordingly, both the covering letter and questionnaire requested that the individual in the household who dealt with issues such as shopping and recycling filled out the survey. However, it is readily acknowledged that the data presented in this and forthcoming chapters is representative of individuals *within* households, not households themselves.

After data collection was complete and coding had been undertaken, a total of 1,265 usable questionnaire responses were received, representing a return rate of 79 per cent (the percentage of questionnaires returned from households who agreed to participate in the research), from a response rate of 59 per cent (the total number of selected households who participated). These excellent returns mean that the results which now follow can be treated with a good level of confidence. Accordingly, a discussion on the social composition of the sample will be undertaken in Chapter 7, specifically related to the composition of lifestyle groups. It is considered that the high return rate negates the necessity to provide a detailed breakdown of the sample at this stage.

Reported Behaviour

This section will examine the responses to Question 1 of the survey, which asked respondents to state how often they undertook a series of 40 environmental actions. These will be examined in the context of the four behaviours that were measured in the questionnaire

Energy Saving

Figures 6.2 and 6.3 illustrate the frequency with which individuals undertook each of the energy-saving behaviours. With regard to the habitual behaviours in Figure 6.2 (the first six items), there is significant variability. The most regular behaviours were reducing heating in unoccupied rooms, switching off lights in such rooms and waiting for a full load before using the washing machine. Over three quarters of individuals undertook these actions 'always' or 'usually'. These are in some contrast to the other three variables (reducing hot water temperature, putting on more clothing and keeping the heating low), which had more individuals who stated that they 'never' or 'rarely' undertook such behaviours. The crucial difference between these two types of items appears to be that the latter are related to the level of comfort one might experience and a consequent reduction in such comfort if these actions were followed. The former relate more to saving energy that otherwise would be wasted, with no benefit to the user. This divergence accords well to the literature on energy saving (see Chapter 5) that has found a distinct difference between energy saving for a clear economic benefit to the individual and energy saving that impacts on comfort.

The final two items in Figure 6.2 and the two binary items reported in Figure 6.3 present data relate to either less regular forms of activity or physical characteristics of a respondent's home. Starting with *energy* items 7 and 8 in Figure 6.2, which relate to forms of consumption and represent less regular forms of decision-making, there is a notable difference in frequency between these types of behaviour and more habitual activities. Only a quarter of the sample regularly looked to replace light bulbs with energy efficient bulbs, although looking for energy efficient appliances was more regularly undertaken, with the vast majority

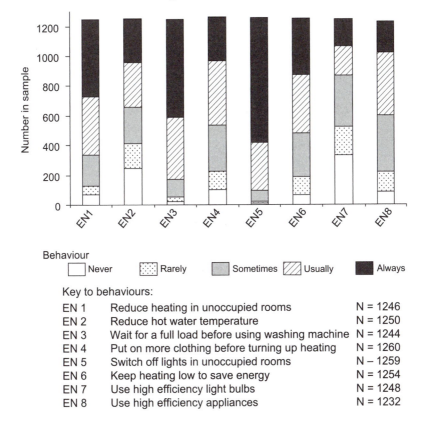

Behaviour

☐ Never ▦ Rarely ▨ Sometimes ▧ Usually ■ Always

Key to behaviours:

EN 1	Reduce heating in unoccupied rooms	N = 1246
EN 2	Reduce hot water temperature	N = 1250
EN 3	Wait for a full load before using washing machine	N = 1244
EN 4	Put on more clothing before turning up heating	N = 1260
EN 5	Switch off lights in unoccupied rooms	N – 1259
EN 6	Keep heating low to save energy	N = 1254
EN 7	Use high efficiency light bulbs	N = 1248
EN 8	Use high efficiency appliances	N = 1232

Figure 6.2 Energy-saving behaviour in the survey

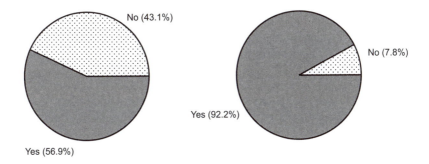

Energy efficient boiler or Economy 7

No (43.1%)

Yes (56.9%)

Insulation

No (7.8%)

Yes (92.2%)

Figure 6.3 Energy-efficiency behaviour in the survey

of the sample looking at least 'sometimes' for these goods, probably on an item-for-item basis.

Figure 6.3 provides data relating to two physical characteristics of the home environment: the heating system and insulation. Over 40 per cent of the sample did not have an energy efficient boiler. This is an important but not surprising result. Central heating systems are regressively expensive and it is likely that a reliable boiler will not be replaced with an energy efficient one until such time as it needs replacing in any case. However, well over half of the sample did have such a boiler, be that through circumstance (many rented properties tend to have storage heaters) or through a conscious decision to enhance energy efficiency. Part of this may be accounted for by the inclusion of 'Economy 7' in the question, which is a storage heating system that uses energy at night to store heat for the following day. Although this cuts costs for consumers, the inefficiencies of the actual heating system may mean that extra use of back-ups during the day could negate any energy saved during the night period. The second pie chart in Figure 6.3 is apparently very positive, but caution needs to be exercised in this instance because insulation was operationalised in the questionnaire as any of the following: walls (cavity), roof, windows or doors. Accordingly, that over 90% reported having such insulation is unsurprising, given that all modern homes would be built with such insulation and local authorities have for many years provided grants for the installation of lofts and cavity wall insulation. In terms of physical measures to save energy, the sample therefore showed variability in the amount of measures undertaken, from a vast majority who had insulation, to a minority who chose to purchase energy efficient light bulbs regularly. It is likely that these different behaviours are the result of alternative antecedents, reflected in both the situational characteristics of a building and the psychological perceptions of expending money for energy efficient light bulbs.

Water Conservation

Figures 6.4 and 6.5 provide information relating to the ten water conservation activities that were examined in the questionnaire. Once again, the analysis of these behaviours can be split according to habitual activities and physical alterations to the home environment, In terms of the habitual behaviours, items 7 and 8 in Figure 6.4 (relating to those who had access to a garden only) show that the majority of individuals with gardens either did not use a sprinkler, or if so, did so very sparingly. Conversely, only a sixth of the sample bought plants that reduced water use. Given the recent hot summers and, in the early 1990s, the proliferation of hosepipe and sprinkler bans, along with licences for the latter, the use of sprinklers may have become less common. Indeed, the promotion of water meters in the south west of England may have led to conscious economic decisions of householders to reduce the use of 'wasteful' forms of water usage such as sprinkler systems. However, the use of plants that do not use as much water as conventional garden foliage is probably far less common, since as a nation of gardeners, Britons are keen to demonstrate their gardening skills by the use of different plants, flowers and high levels of ground cover, all requiring

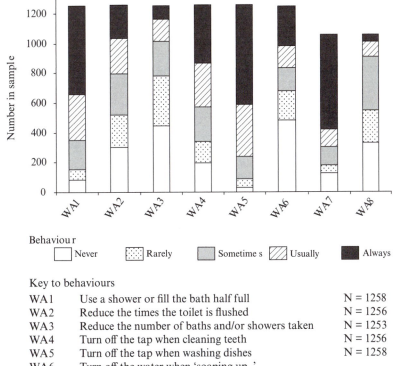

Key to behaviours

WA1	Use a shower or fill the bath half full	N = 1258
WA2	Reduce the times the toilet is flushed	N = 1256
WA3	Reduce the number of baths and/or showers taken	N = 1253
WA4	Turn off the tap when cleaning teeth	N = 1256
WA5	Turn off the tap when washing dishes	N = 1258
WA6	Turn off the water when 'soaping up ' in the bath or shower	N = 1248
WA7 *	Use a sprinkler sparingly , if at all, in the garden	N = 1055
WA8 *	Buying plants that require less water	N = 1054

* Note: Only those with access to a garden could answer these items.

Figure 6.4   Water conservation behaviour in the survey

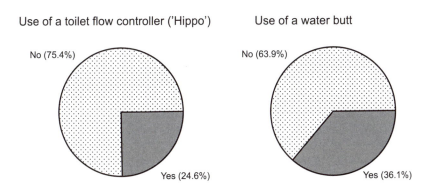

Figure 6.5   Use of water saving devices in the survey

considerable amounts of water. Nonetheless, the drought of 2006 has witnessed further uptake of less water-intensive plant species, especially in the south east of England.

The remaining 'indoor' behaviours show considerable variability (Figure 6.4, items 1–6). There are again two distinct groupings according to frequency. A majority of individuals always used a shower or less water in the bath and turned off the tap when washing dishes and cleaning teeth. Conversely, a majority never or rarely took fewer baths or showers to save water and over a third never or rarely flushed the toilet less to save water. This is also reflected in the frequency that individuals turned off the water to 'soap up' in the shower or bath. As with energy-saving behaviour, indoor water conservation activities appear to be split into those that reduce 'waste' water, such as turning off taps when washing dishes and those actions that require some personal sacrifice, perhaps in perceived cleanliness. The former may be influenced heavily by the growing popularity of water meters, yet it is ironic that most household water is used flushing the toilet or washing.

Turing to the physical alterations within the home environment to conserve water (Figure 6.5), three-quarters did not have a water saving device in their toilet cistern and almost two thirds of households with a garden did not use a water butt. Therefore, unlike the case of insulation above, it appears that what are fairly simple alterations within the home have not been taken by most households. The use of such devices is evidently different from the use of, say, insulation, since water butts and hippos can be removed and installed very simply. Nonetheless, it is instructive that the majority of households do not use such devices.

Green Consumption

Figure 6.6 provides data relating to the consumption habits of the respondents. As highlighted previously in this chapter and demonstrated clearly in Figure 6.6, the diversity in 'green' consumption activities is considerable. The immediate point to note in Figure 6.6 is the stark contrast between items 9 and 10 (using a local food store and buying local produce) and items 1 to 8. Around a third of individuals in the sample reported always or usually using a local food store and purchasing local produce. Evidently, these items do raise pertinent questions, not least whether a local store is a corner shop, butcher, baker, or a Tesco supermarket. The term 'local' is heavily contested (Winter 2003) and has been used in oppositional terms to the sprawling commercial empires of major retailers such as Tesco, Sainsbury's, Asda-Walmart and Morrison's. The finding in this survey that so many reported the use of a local food store is, quite simply, in contrast to the large numbers that patronise supermarkets in the UK and therefore this item may yield very little useful data on the use of 'local' shops. However, if 'local' is conceptualised in an alternative manner, then individuals answering this question may simply be reporting that they *only* drive a few miles to shop, regardless of what retail outlet is being used. Indeed, there may be some dispute over what constitutes 'local produce'. Supermarkets and independent shops both provide 'local' products, with scales of locality being stretched to incorporate whole regions. These contestations of the local present interesting issues for discussion pertaining to the role of both consumer and producer definitions of what constitutes local. Nonetheless, taken at face value, there does appear to be a majority of the sample who at least sometimes shopped and bought locally and the reader should consult Gilg et al. (2005) for more information relating to local food networks.

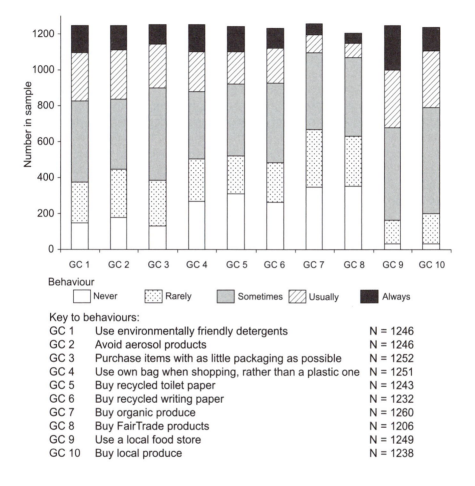

Key to behaviours:

GC 1	Use environmentally friendly detergents	N = 1246
GC 2	Avoid aerosol products	N = 1246
GC 3	Purchase items with as little packaging as possible	N = 1252
GC 4	Use own bag when shopping, rather than a plastic one	N = 1251
GC 5	Buy recycled toilet paper	N = 1243
GC 6	Buy recycled writing paper	N = 1232
GC 7	Buy organic produce	N = 1260
GC 8	Buy FairTrade products	N = 1206
GC 9	Use a local food store	N = 1249
GC 10	Buy local produce	N = 1238

Figure 6.6   Green consumption behaviours in the survey

The remaining items in Figure 6.6 are proportionally quite similar. Few individuals usually purchase organic products and Fair Trade produce. Indeed, no more than a quarter of individuals ever undertook any of the other behaviours regularly. The large amount of individuals who reported that they 'sometimes' undertook a behaviour raises questions over whether purchase motivations are governed by environmental concerns or, for example, price. For example, people may sometimes buy recycled toilet paper if it is cheaper or on offer, but not regularly. Accordingly, with the exception of a tendency to buy local produce (which has definitional and conceptual issues associated with it), the level of green consumer behaviour in the sample was sporadic to say the least. This ranged the different conceptual boundaries that had been placed on the items, from avoiding environmentally damaging products to waste reduction and buying recycled produce.

Waste Management

Figure 6.7 provides data relating to the frequency with which individuals in the sample undertook a range of waste-related behaviours. The major point that can be drawn from these data is the general support for a wide range of the activities measured. Large majorities recycled all of the items mentioned with regularity. This reinforces the notion that recycling is normative behaviour, although there is some variation, with significant numbers reporting that they never recycled cans or plastic bottles. In terms of re-use, the power of the charity shop sector can clearly be seen, along with the regular re-use of paper within the household. Nonetheless, donation of furniture to charity was still less frequent. Overall, however, re-use was well established amongst the sample. These findings lend more evidence to arguments presented elsewhere (e.g. Gilg et al. 2005) that recycling behaviour is a very distinctive activity, which is heavily dependent on local authority service provision. Recycling involves a structured programme of behaviour, which is mostly undertaken at the kerbside and is therefore governed by both social norms but also, with increasing importance in the UK, the enforcement of recycling policies that compel householders to sort their waste before collection.

In stark contrast to this positive situation, the final two items in Figure 6.7 provide cause for concern and demonstrate that structured recycling is significantly different from other forms of waste-related activities. Item 9 demonstrates that a majority never composted their kitchen waste. This is a major area of concern, since 40 per cent of waste being sent to landfill is biodegradable and the European Union's Landfill Directive compels the UK to reduce biodegradable waste significantly by 2015 (see Chapter 3). The government is keen to see composting increase radically over the next decade or so, in order to replace the perceived limit to recycling of other wastes. The figures for composting garden waste are not encouraging either. In the popular media, composting has been portrayed as both dirty and unhygienic. Accordingly, in contrast to the positive situation relating to recycling, composting remains a major challenge for policy-makers seeking to reduce the flow of biodegradable waste into the landfill stream.

Linking Behaviours: Framing Practices

The data presented in Figures 6.2 to 6.7 provide an insight into the current state of environmental action in Britain at the beginning of the twenty-first century. They demonstrate some significant differences between both types of activity (habitual vs. physical) and behavioural domains (within vs. beyond the home).

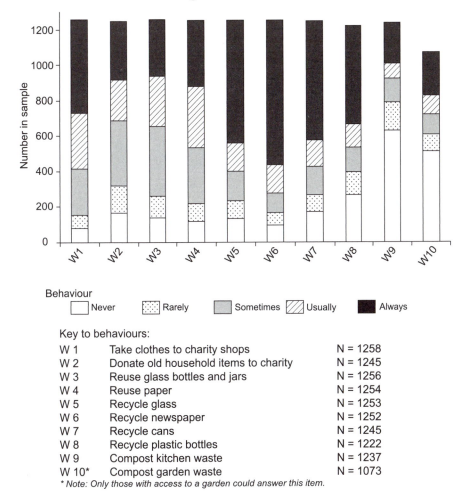

Figure 6.7   Waste management behaviours in the survey

Indeed, these differences may be greater than the apparent (artificial) differences between the four environmental behaviours that we have examined. These are specifically related to environmental problems (such as energy or water use). However, although this has been the dominant vehicle by which both academics and practitioners have chosen to study and promote environmental action, the evidence demonstrated above clearly shows that there are likely to be links between different environmental actions and that these may be reflected in the everyday practices that individuals are engaged in. It is highly improbable that individuals act out environmental practices in a separate behavioural domain than their everyday routines. Accordingly, this section seeks to further address the first empirical objective of the research by examining the links between environmental behaviours and how these relate to everyday practices in and around the home.

From Behaviours to Practices: Linking Environmental Action to
Everyday Practice

The hypothesis that environmental behaviours can be linked and related to forms of everyday practice provides the researcher with an analytical problem. Individuals answering a questionnaire survey treat each item that they answer as an individual question about their behaviour. If the survey team had asked individuals to state how each of the forty activities related to each other, numerous alternatives would have been forthcoming. Indeed, just as respondents may have differing views about how their individual responses might be fitted together, so do researchers. Accordingly, attempting to reliably link behaviours together in an ad hoc manner presents numerous analytical and conceptual problems.

To overcome these difficulties, social science researchers have applied a series of related techniques that can collectively be termed 'factor analysis' to statistically relate items in a questionnaire to each other and, on the basis of these links, reduce the data to a series of 'factors' that are representative of the wider set of data. Accordingly, factor analysis has two major roles, intimately related to the aims of a research project. First, it can be used to identify the links between items in a questionnaire, to classify variables into a series of groups which can be related to concepts or constructs. Second, it can be utilised as a core proactive analytical tool, reducing the number of items in a questionnaire needed for analysis and, through the summation of the items in each factor, can be used to create new data scales.

Factor Analysis

Both of these uses for factor analysis are considered to be 'exploratory' in nature. In essence, the researcher poses the question: 'are there any links between my items in a questionnaire and can the number of items be reduced?'. In 'confirmatory' factor analysis, the researcher uses the technique to establish whether, after data reduction is complete, their research has matched or differs from previous studies using similar data. Technically, exploratory factor analysis determines the degree to which questionnaire items can be aggregated into more generalisable variables (or factors) and the extent to which such items can be empirically related and therefore be said to be 'factorially valid' (Bryman and Cramer 1996). Determining on which 'factors' these items 'load' and the extent to which these loadings mirror expectations of how items aggregate enables the researcher to validate their conceptual structure. Indeed, aggregation enables the researcher to meet some of the requirements for multiple regression analysis (such as normality). In the current study, those questionnaire items that were measured on a five-point Likert scale were analysed using 'principal components' factor analysis (Wheeler et al. 2004), a technique for describing the extent to which individuals share variance in their scores on specific items. When factor analysis is initially undertaken, this variance is described in terms of 'factors' or 'components', with the first factor extracted from an analysis accounting for the largest amount of variance. At this stage, there areas many factors as items. In order to reduce these to a meaningful set of factors (to meet the requirements of aggregation), certain factors must be dispensed with and others 'retained'. Retention of factors is undertaken on the basis of the amount of variance explained by factors. Thus, those factors that explain less variance than for a single item are excluded. This is done by examining the 'eigenvalues' that express this variance and excluding any with scores under one. Alternatively, a visual retention technique can be employed, whereby the eigenvalues are plotted as a 'scree plot' and the factors on the flat end of the slope are excluded. Once a set of factors that represent acceptable amounts of variance have been retained, it is necessary to 'rotate' these factors, since initial analysis will have meant that most items loaded on the first factor, but with low correlations. In most cases social scientists utilise 'orthogonal' rotation, since it can be assumed that item scores on one factor are generally unrelated to their scores on another factor. Once rotated and 'sorted' for ease of interpretation, the resulting output from a statistical package represents the number of principal components or 'factors' that represent the shared variance of the items in question. Thus at the bottom of the tables, the 'Variance' represents the eigenvalue for each factor, whilst the '% variance' represents the total variation explained by that factor and is interpreted as a percentage (e.g.0.392 = 39.2%). Factors are identified by the shading of the correlation coefficients and conform to the degree to which each item 'loads' on that factor, the higher the coefficient, the higher the loading. Thus within the theoretical constraints placed upon the researcher, the questionnaire items can be examined for common variance in order to assess the degree to which questionnaire items load and enable the researcher to proceed with more complex analyses in the knowledge that the aggregated data has been constructed using an established technique.

Although factor analysis immediately appears to be a somewhat complex technique, it is potentially useful for identifying links between items in a questionnaire and on the basis of the results, identifying types of activity that individuals are undertaking. Evidently, the selection of factors to retain is based on the judgment of the researcher and this can have a significant impact on the meaning and interpretation of the results. Nonetheless, if the rules outlined above are strictly applied, the factorial solution should represent statistically significant links between items in a questionnaire.

Aggregating the Survey Data

Table 6.7 presents the results of the principal components factor analysis undertaken on 36 of the 40 behavioural items. Because two energy and two water saving items were measured on binary scales, these could not be placed into the analysis. The items from the questionnaire are given in the second column of the table and are listed in the order they were outputted from the factor analysis. The first column of the table provides a name for the factor, which was subjectively assigned by the research team and will be examined below. The variance explained by each factor is given in the third column and, as can be seen, all of the factors have eigenvalues of well over 1.0. Indeed, although the cumulative variance explained by the factor solution is only 36 per cent, this is spread relatively evenly

Table 6.7   Behavioural factors derived from Principal Component Analysis

Factor	Variables included	Item code	Variance (eigen-value)	Per cent variance	Cronbach's Alpha
Purchase decisions	Use high efficiency light bulbs	EN7	4.4	13.3%	.83
	Use high efficiency appliances	EN8			
	Buy organic produce	GC7			
	Buy Fair Trade products	GC8			
	Avoid aerosol products	GC2			
	Compost garden waste	W10			
	Compost kitchen waste	W9			
	Use environmentally-friendly detergents	GC1 W4			
	Re-use paper	W3			
	Re-use glass bottles and jars	GC6			
	Buy recycled writing paper	GC5			
	Buy recycled toilet paper	GC10			
	Buy local produce	GC9			
	Use a local store	GC4			
	Use own bag when shopping, rather than a plastic one	GC3 WA8			
	Purchase items with as little packaging as possible				
	Buying plants that require less water				
Habits	Turn off the tap when washing dishes	WA5 WA6	3.9	11.7%	.81
	Turn off the water when 'soaping up' in the bath or shower	WA3 WA2			
	Reduce the number of baths and/or showers taken	WA4 WA1			
	Reduce the times the toilet is flushed	EN1 EN2			
	Turn off the tap when cleaning teeth	EN6 EN3			
	Use a shower or fill the bath half full	EN4 EN5			
	Reduce heating in unoccupied rooms	WA7			
	Reduce hot water temperature				
	Keep heating low to save energy				
	Wait fir a full load before using washing machine				
	Put on more clothing before turning up heating				
	Switch off lights in unoccupied rooms				
	Use a sprinkler sparingly, if at all, in the garden				

Table 6.7 cont'd

Factor	Variables included	Item code	Variance (eigen-value)	Per cent variance	Cronbach's Alpha
Recycling	Recycle glass	W5	3.5	10.5%	.78
	Recycle newspaper	W6			
	Recycle cans	W7			
	Recycle plastic bottles	W8			
	Donate old household items to	W2			
	charity	W1			
	Take clothes to charity shops				
Total variance				36%	

Notes

The final three columns of the table relate to each of the three factors.
The table only includes items measured on Likert scales.
The order of variables relates to the output from the factor analysis.

across all three factors, indicating that all three factors are equally valid. The final column presents a Cronbach Alpha coefficient, which is a measure of internal consistency for each factor. This measure is used when the researcher seeks to use the items in a given factor as a new data scale. This can be undertaken by simply summing the scores of each item in the factor to create a new measure. To be considered internally consistent, a Cronbach Alpha normally has to exceed 0.7. Table 6.7 demonstrates that all three factors have items that would provide internally consistent scales for further analysis.

Environmental Practices: Purchase Decisions, Habits and Recycling

The factor analysis revealed three types of behaviours that comprised distinct statistical categories. The largest factor was termed 'purchase decisions'. This factor was the most complex and at first glance appears to present a somewhat muddled picture. However, some important commonalities are apparent in this factor. Primarily, the factor represents a set of activities that are all related to consumption. It contains all of the green consumption activities that were in the initial conceptualisation (see Figure 6.6), but also contains other purchase-related activities that had previously been considered under different headings. First, the purchase of energy-saving light bulbs and energy efficient appliances indicates that energy-saving behaviours can be considered within a consumption context. Such decisions are fundamentally related to spaces of consumption rather than the home environment. Second, the inclusion of a water conservation item (using plants that need less water) indicates that decisions to reduce water usage are also practiced within a consumption context. Third and perhaps most surprisingly, the inclusion of both composting items (kitchen and garden waste) indicates that decisions relating to waste management are also practiced outside the home in consumption spaces. These two activities are evidently ones that require some form of investment in the form of composting equipment, wormeries and home storage containers and therefore necessitate a specific purchase-related decision.

The purchase decisions factor also demonstrates that environmental practices are intimately linked in terms of the connections that individuals make between alternative behavioural settings. For example, the inclusion of the two waste re-use items (glass and paper) indicate that the decision to undertake these activities is part of the (non-)consumption process, whereby individuals attempt to reduce their consumption, illustrated by the inclusion of items that emphasise other waste reduction activities (not using plastic bags and looking for less packaging).

Accordingly, purchase decisions represent both a series of direct consumptive practices, related to green consumption as it has traditionally been conceptualised, but also other behaviours, such as energy, water and waste management. This factor also demonstrates the connections between consumption and other forms of environmental practice and therefore represents indirect decisions relating to consumption that are acted out within both consumptive and home environments.

The second factor in Table 6.7 has been termed 'habits' and relates to a series of items in the questionnaire that measured habitual behaviours, focused around energy and water saving activities. The behaviours are all practiced within the home and relate to everyday practices and habits that are undertaken with significant regularity as part of daily routines. The factor contains items that require little or no sacrifice, such as turning off the tap when washing dishes, to more significant changes to everyday practice, such as reducing the number of times the toilet is flushed and using less water when showering. Although the latter items are still habitual in nature, they require a greater level of commitment and relate to notions of convenience, comfort and the degree to which environmental priorities take precedence over previously constructed attitudes towards personal hygiene and 'cleanliness'. Nonetheless, it is significant that all of these activities relate directly to energy and water conservation and therefore present a second distinct set of environmental practices.

The third factor is the most distinctive and has simply been termed 'recycling'. As noted earlier in this chapter, the nature of recycling behaviour within the UK means that it is a significantly different type of activity from other forms of environmental action. The factor containing all of the recycling items corroborates this assertion and highlights recycling behaviour as a distinctive post-consumption activity within the home. However, as noted earlier, the importance of the charity sector as a vehicle for the re-use of household items is also significant. The inclusion of these two items in this factor demonstrates the significance of charity shops and collections to the waste sector.

Environmental Practice as Everyday Practice?

The analysis provided in Table 6.7 and in the previous section clearly demonstrates the analytical links between environmental behaviours. These can legitimately be translated into conceptual links between environmental activities. Such connections clearly demonstrate that three sets of environmental practices are evident in contemporary British society. The first relates to purchase decisions, incorporating a wide range of decisions that individuals make relating to a diverse set of environmental dilemmas. These span energy concerns, water conservation, environmental pollution, ethical consumption and waste management. These practices are acted out within consumption spaces and therefore represent environmental action beyond the home environment. Second and in direct contrast to these consumption behaviours, habitual practices are undertaken within the home and relate to energy and water conservation practices, intimately linked to daily practices. Third, recycling behaviour represents a highly-structured and regulated set of practices for dealing with post-consumer waste within and, in the case of the charity sector, beyond the home. Accordingly, these three sets of practices represent a new way of examining environmental behaviour, relating activities to each other and understanding the connected nature of environmental action.

However, the data presented in Table 6.7 also represents a further innovation in the way in which both academics and practitioners conceptualise environmental action. The three factors identified relate to three major spaces in which practices are acted out in modern society. In the first instance, the purchase decisions factor pertains directly to spaces of consumption, within which environmental practices will play a significant part, embedding decisions relating to environmental issues within a myriad of other priorities concerning products and brands. Factor 1 in Table 6.7 demonstrates that environmental practices need to be situated within existing spheres of consumption, rather than being separated from them. Similarly, Factor 2 demonstrates the need to view environmental practices in the home as part of everyday lifestyle choices. As emphasised above, some choices to help the environment can be more easily embedded into everyday practice than others, given that the environment as a priority competes with other concepts such as comfort, convenience and hygiene. Nonetheless, habitual practices to help the environment need to be seen as part of everyday decisions concerning routines and habits that we take for granted. Finally, Factor 3 may be representative of a completely new set of everyday practices which are directly motivated by environmental concerns, thus distinguishing recycling from the other two sets of practices. Recycling, as distinct and 'novel' practices, have become embedded into everyday routines in the home and therefore need to be regarded as distinct from other forms of waste management behaviour. Accordingly, an argument can made that links environmental practices to everyday lifestyle choices, which are undertaken in the same domain, rather than representing fundamentally different lifestyle decisions.

Conclusion: Contemporary Environmental Practice

The results reported in this chapter indicate that current environmental practices are undertaken with varying frequency, but the overall picture is positive in many respects. Recycling was undertaken frequently by respondents and many engaged with habitual actions to save energy and water. More concerning was the low uptake of green consumption activities and composting behaviour. The initial analysis of the behavioural data therefore points to consumption as being one area where more research and practical engagement is required. This is partly because the very nature of consumption is contested and encompasses such a wide array of activities, from ethical to ecological forms of consumption.

A key theme of this text has thus far been the compartmentalisation of both academic and practical work into the promotion of environmental action, focused around environmental problems. Specific research projects or environmental campaigns have highlighted the need to encourage 'energy saving' or water conservation'. The findings from the ESRC project reported above would clearly indicate that academics and policy-makers need to shift their emphasis from problems to practices. Environmental campaigns that have emphasised an approach highlighting environmental problems and recommending a series of actions are likely to have 'skimmed' across a series of behavioural domains that represent several forms of environmental practises. The research reported in this chapter suggests that attention needs to be turned towards focusing on three major 'spaces' of environmental practice, related closely to everyday practice. Focusing on environmental aspects of consumption, everyday behaviours in the home and recycling could yield more positive results for practitioners who seek to encourage environmental action. By emphasising how individuals can change their behaviour within the same spaces that their existing practices are performed, the shift towards an environmental lifestyle is likely to be greeted with less suspicion and trepidation. The current stereotypical view of the environmentalist as a frugal individual who has a 'miserable' lifestyle, with little interest in consumption and living 'in the dark' must be shifted towards an emphasis on free consumer choices concerning 'different' forms of consumption and lifestyle practices, rather than 'less'. This topic will be developed further in the forthcoming chapters.

These data answer the first objective of the research, which was to identify how different environmental practices relate to each other and to everyday practices. The next objective attempted to problematise the notion that individuals act in similar ways and react to environmental messages with like-minded responses. Just as we have critiqued the notion that environmental behaviour can be seen as conceptually distinct from other forms of behaviour, the second objective sought to understand the extent to which environmental action can be classified into different segments and, on the basis of these segments, whether lifestyle groups can be identified to further understand how environmental practices are undertaken by different parts of the population.

To illustrate how these and the other objectives for the study link into the overall structure of this book and the processes of data analysis used, Figure 6.8 provides an analytical structure for the following three chapters. As can be seen, the progression of data analysis moves from descriptive to explanatory, with a move from a more academic focus to a policy-orientated perspective. Chapter 6 has explored the nature of environmental practice and how it is framed. Chapters 7 and 8 continue to utilise these quantitative data through an exploration of sustainable lifestyles and the influences which act on differing levels of environmental commitment. Chapter 9 seeks to apply both these data and those from further qualitative research in an exploration of the policies that might be developed for promoting behaviour change.

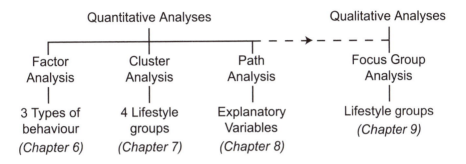

Figure 6.8   Analytical structure of the book

Chapter 7

Sustainable Lifestyles

Introduction: Segmenting for Sustainability

At the end of the last chapter, the assertion was made that even if environmental behaviours could be classified into a series of activities that reflected everyday practices in and around the home, the assumption that behaviour change policies could simply be adjusted to nuance messages towards these types of activities was problematic. Even if a set of key influences on such practices could be identified (Figure 4.3), this would be grounded in the assumption that such influences stretched across a range of social and lifestyle groups in society. Accordingly, before examining the influences on these alternative practices, a key question has to be posed: is there good evidence to suggest that identifying social or lifestyle groups, each with their own unique set of attributes, would assist in the understanding and promotion of greater levels of environmental action?

Before answering this question in detail, let us reflect on one illustrative example unrelated to the environmental field. The commercial sector has at its heart the principle that products and services can be targeted at specific individuals, who are more commonly referred to as 'market segments'. Market research agencies and opinion polling companies such as NOP, MORI and UGOV have all sought to create market profiles for a range of goods and services, as well as providing advice to political parties on how to target their policies at specific segments in the population. Within the United Kingdom, General Elections are often the time when the crudities of such market segmentation are most evident, with specific voting groups being labelled variably as 'Sierra Man', 'White Van Man' and 'Mondeo Man'. These seemingly crude and playful terms have their origins within market research, which uses a number of social and attitudinal measures to assign individuals to particular segments according to their attributes. Commercially, the use of loyalty cards by supermarkets has further developed notions of market segmentation, with the targeting of specific offers, goods and services to individuals who display, by virtue of their shopping behaviour, particular trends in their consumption. Virtually any product can be used as an example, but the development by large supermarket chains of economy label products is a useful example illustrating the use of market segmentation and brand identification. The use of the 'value' labels by Tesco in the early 1990s was designed specifically to relate to householders on low incomes who needed to budget, whilst still having access to a series of valued products. At the same time, the splitting of other own brands into categories such as 'Tesco's Finest' distinguished these products from the normal own brands, offering an alternative to branded goods that may have been regarded as of better quality. The promotion and advertising of these products relies heavily on relationship marketing, using a range of messages conveyed both visually and linguistically to relate to specific groups in society. Perhaps the most recent incarnation of this type of relationship marketing has been the recent proliferation of 'Fairtrade' produce in mainstream supermarkets, which has emerged from the identification by supermarkets of a distinctive market in 'ethical' goods.

Without a doubt, such techniques have been extremely successful, enabling large commercial enterprises to develop relationship and brand marketing for groups of people who may have previously been 'missed' by conventional broad brush advertising and promotional campaigns. We will revisit these ideas in Chapter 9 with specific reference to branding and the promotion of environmental action, but suffice it to state that marketing strategies that have employed the principles of market segmentation have been able to accurately and cost-effectively target products to new groups of individuals who have yet to engage with their product.

Let us now return to the question posed at the start of this section: is segmenting for sustainability a useful concept and what evidence is there that it can contribute to behaviour change policy? One of the hallmarks of behaviour change policy during the 1990s and in the early part of the twenty-first century has been the assertion that 'the public' comprises a homogeneous group of individuals who will positively react to policies that seek to change their lifestyles. A raft of research during the last five years has demonstrated that this assumption is both flawed and growing evidence suggests that environmental practices are significantly nuanced according to a series of social or lifestyle groupings (Darnton 2004a; 2004b). The hypothesis that environmental practices vary according to specific lifestyle segments derives from previous work (e.g. McKenzie-Mohr 2000) that has begun to promote social marketing techniques as a means by which to increase environmental behaviour. McKenzie-Mohr (2000) has argued that social marketing seeks to recognise the implicit barriers that specific groups in society face in undertaking particular behaviours and therefore aims to work with these particular groups to remove such barriers. Within the most basic model of social marketing, Darnton and Sharp (2006a; 2006b) have identified a large number of segmentation models that have been used by both practitioners and academics to understand how social and lifestyle groups can be used to identify the influences on environmental action and can be used for changing behaviour.

Darnton and Sharp's (2006a; 2006b) study examined a range of segmentation models. Segmentation as a technique seeks to utilise a range of quantitative data on individuals to identify similarities and link individuals with the closest set of attributes, eventually forming a series of groups, or segments. Within the field of environmental action, two types of segmentation can be identified: those based on socio-economic characteristics and those based on a range of behavioural and attitudinal data. In the first instance, an implicit assumption is made that socio-economic status is either a 'cause' or at least a very good proxy for environmental action. Researchers may use conventional measures of socio-economic status (or class) such as the A, B, C1, C2, D and E categorisation (Darnton and Sharp

2006a) or spatially defined measures of social grouping, such as Census data. In some cases, spatial data relating to specific environmental performance (such as recycling rates) can be linked to socio-economic data and a series of spatially-refined socio-economic groups can be identified. Although these methods of segmentation present a relatively simple means by which to divide-up population and an effective methodology for implementing policy on a geographical scale, they are grounded in the assumption that social class is a reasonably accurate proxy for environmental action. As Darnton (2004a; 2004b) notes, the links between socio-demographics and environmental action are at best equivocal.

Accordingly, the second and more common type of segmentation utilised, particularly by academics, is focussed around the use of sample survey data collected on environmental action and attitudes towards the environment. Darnton and Sharp (2006a) have noted the wide range of segmentation techniques used in these studies. The most notable difference relates to the variables that researchers have used to segment their samples. For example, researchers have segmented according to behavioural characteristics, attitudinal variables or a combination of both. These mostly vary with the types of research project being undertaken and whether the researcher is interested in constructing a profile based on reported behaviour or a more holistic perspective including a wider range of personality and perceptual attributes.

The importance of segmentation for sustainable development has recently been emphasised by recent moves within DEFRA to utilise social marketing and segmentation for behaviour change. However, as Darnton and Sharp (2006a) have highlighted, the problems faced by policy-makers who seek to simply apply certain models or approaches are that the subjective nature of segmentation (implicit in the process of a researcher identifying key groups from an analysis) and the plethora of models available makes a 'quick fix' both impractical and ineffective. Accordingly, the current state of understanding is that whilst segmentation may be useful, the most pressing issue is how to define segments and effectively target them. One proposal for overcoming this problem is provided in Chapter 9. However, this chapter will focus on highlighting the potential for segmentation to provide a basis for more clearly understanding environmental action in and around the home and how environmental practices may be represented by specific lifestyle groups (see Figure 6.8). The chapter firstly outlines the basis on which the data from the ESRC research (introduced in Chapter 6) were segmented and then examines the attributes of the clusters in terms of social and environmental values, situational characteristics and psychological factors.

Segmenting Environmental Practices: Cluster Analysis

It is helpful to examine the basis of the segmentation that was used for the ESRC study within the context of the wider social scientific application of segmentation techniques. The most common form of segmentation has utilised a statistical technique termed 'cluster analysis'. This technique uses scores on a questionnaire to construct a series of groups, or clusters, based on the 'similarity' of individuals to each other (Wheeler et al. 2004). Within the ESRC research, the 36 items measured on a frequency scale were placed into what is termed a Hierarchical Cluster Analysis. The term hierarchical refers to the process by which all individuals in an analysis (1,265 in the case of this research), which are progressively joined together until they are all in the same group. At the beginning of the process, the scores of the first individual in the data set on all 36 items are compared to all other individuals (or cases), the case with the most similar scores is 'paired' to the first case. The combined scores of this new 'cluster' are then compared to all other cases and then joined to the most similar case or cluster. The pairing process proceeds until just one cluster containing all 1,265 cases is present. The pairing process can be represented graphically as a 'dendrogram', demonstrating where new clusters have been formed and identifying instances where major clusters have joined.

Cluster analysis is a statistical technique for pairing cases according to their similarity. It is the researcher's role to examine the pairing process using the dendrogram and to make key decisions regarding the number of clusters which seem to be significant. This can often be problematic, since a wide range of clusters can suddenly be condensed into just one or two clusters. Accordingly, interpretation of cluster analyses must take into account the subjective nature of deciding how many clusters are significant. The most effective means of deciding how many clusters to retain for analysis is to examine where the largest breaks in the dendrogram can be found and, even if clusters are relatively uneven in number, using these for analysis. In the case of the ESRC research a total of four definitive clusters emerged from the cluster analysis, although the fourth cluster was very small.

It is worth noting that only behavioural items were placed into the cluster analysis. None of the situational, psychological or social / environmental value items were examined, mainly because the research was primarily interested in predicting environmental behaviour. The addition of other variables would have meant that the variables being used to predict behaviour (Chapter 8) would also have been used partly as the basis for the groups being examined. The following analyses therefore focus on the attributes of the four clusters identified in the research in terms of their reported environmental behaviour, situational variables, social and environmental values and psychological variables. This will serve to construct a profile of each cluster and examine the extent to which these are representative of lifestyle groups. Out of necessity, the following is a summary of the findings, more detailed analysis of which can be obtained from Barr and Gilg (2006), Barr et al. (2005a; 2005b) and Gilg and Barr (2005a; 2005b; 2006) and which provided the basis for the forthcoming sections.

Lifestyle Groups 1: Environmental Practices

The four clusters were comprised of the following numbers of individuals from the sample of 1,265:

- Cluster 1 (N 294);
- Cluster 2 (N 412);
- Cluster 3 (N 505);
- Cluster 4 (N 43).

Although Cluster 4 was very small, its distinctive nature throughout the segmentation process was noted from the dendrogram. Figures 7.1 to 7.4 provide a visual representation of the frequency with which individuals undertook the range of environmental actions measured. These figures have been arranged according to the factorial measures in Table 6.7. As can be seen from Figure 7.1 to 7.4, each cluster has been given a title, which is purely subjective, but which seeks to provide a brief indicative description of the behavioural properties of the cluster. We will now look at these in turn.

Committed Environmentalists

Cluster 1 was classified as a group of 'Committed Environmentalists' (Figure 7.1). These individuals regularly engaged with the majority of the behaviours in all of the three factors (purchase, habitual and recycling) that represented the three distinct environmental practices examined in Chapter 6. Starting with purchase decisions, there was generally a high commitment to the range of practices within this factor. The most regular practices related to composting and reuse behaviour, with over 60 per cent of this cluster always composting kitchen and garden waste and around 50 per cent of individuals reporting that they reused paper and glass. Other activities undertaken with significant regularity were the avoidance of plastic shopping bags, looking for items with less packaging, the purchase of detergents without toxic agents, avoidance of aerosols, purchase of local food and the use of local food shops. The least common activities related to the purchase of fairly traded goods, organic produce and high efficiency light bulbs, although well over 50 per cent of respondents reported undertaking each of these behaviours 'sometimes'.

The analysis of this cluster suggests that Committed Environmentalists thought carefully about their consumption choices, particularly in relation to the negative environmental impacts of consumptive activities, related to aspects of environmental pollution and waste management. Other activities related to ethical consumption and organic products were less popular, although there was apparent support for local food. These results indicate that this group were a highly committed and motivated set of individuals, willing to make discerning choices about consumption, specifically searching out key products. Of particular significance in this cluster was the vast majority who stated that they always or usually composted waste, in sharp contrast to the general picture given in Figure 6.7. This reinforces the notion that this group was 'committed', given the marginal nature of composting demonstrated in the overall results.

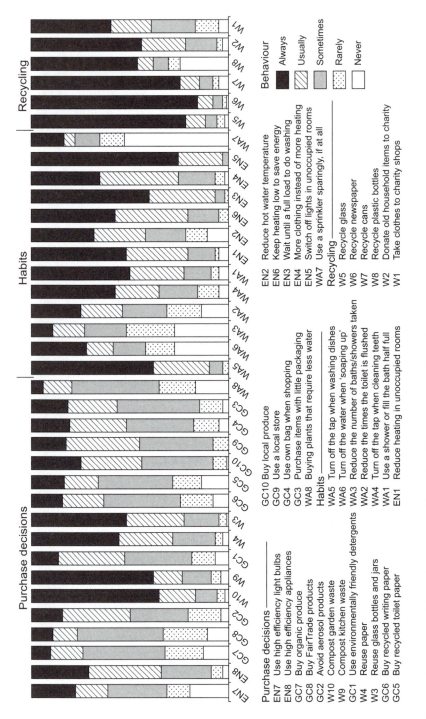

Figure 7.1 Reported behaviour for the Committed Environmentalists cluster

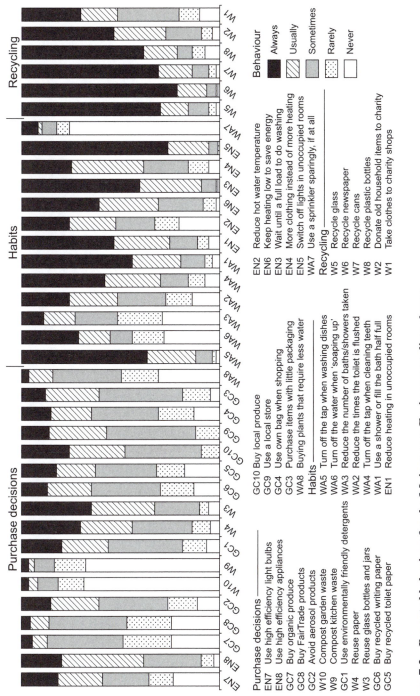

Figure 7.2 Reported behaviour for the Mainstream Environmentalists cluster

Purchase decisions
EN7 Use high efficiency light bulbs
EN8 Use high efficiency appliances
GC7 Buy organic produce
GC8 Buy FairTrade products
GC2 Avoid aerosol products
W10 Compost garden waste
W9 Compost kitchen waste
GC1 Use environmentally friendly detergents
W4 Reuse paper
W3 Reuse glass bottles and jars
GC6 Buy recycled writing paper
GC5 Buy recycled toilet paper

GC10 Buy local produce
GC9 Use a local store
GC4 Use own bag when shopping
GC3 Purchase items with little packaging
WA8 Buying plants that require less water

Habits
WA5 Turn off the tap when washing dishes
WA6 Turn off the water when 'soaping up'
WA3 Reduce the number of baths/showers taken
WA2 Reduce the times the toilet is flushed
WA4 Turn off the tap when cleaning teeth
WA1 Use a shower or fill the bath half full
EN1 Reduce heating in unoccupied rooms

EN2 Reduce hot water temperature
EN6 Keep heating low to save energy
EN3 Wait until a full load to do washing
EN4 More clothing instead of more heating
EN5 Switch off lights in unoccupied rooms
WA7 Use a sprinkler sparingly, if at all

Recycling
W5 Recycle glass
W6 Recycle newspaper
W7 Recycle cans
W8 Recycle plastic bottles
W2 Donate old household items to charity
W1 Take clothes to charity shops

Behaviour
Always
Usually
Sometimes
Rarely
Never

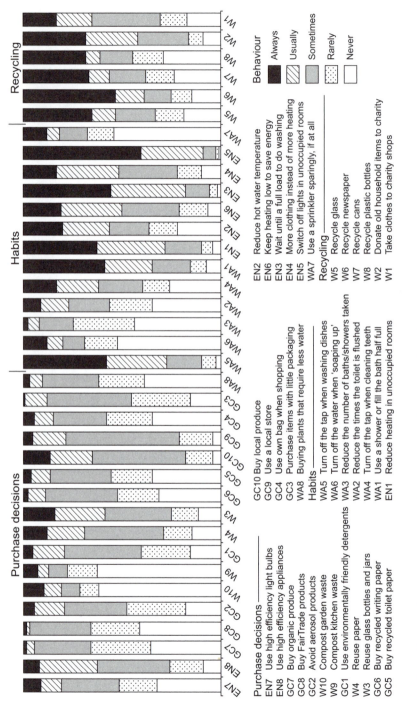

Purchase decisions
EN7 Use high efficiency light bulbs
EN8 Use high efficiency appliances
GC7 Buy organic produce
GC8 Buy FairTrade products
GC2 Avoid aerosol products
W10 Compost garden waste
W9 Compost kitchen waste
GC1 Use environmentally friendly detergents
W4 Reuse paper
W3 Reuse glass bottles and jars
GC6 Buy recycled writing paper
GC5 Buy recycled toilet paper

GC10 Buy local produce
GC9 Use a local store
GC4 Use own bag when shopping
GC3 Purchase items with little packaging
WA8 Buying plants that require less water

Habits
WA5 Turn off the tap when washing dishes
WA6 Turn off the water when 'soaping up'
WA3 Reduce the number of baths/showers taken
WA2 Reduce the times the toilet is flushed
WA4 Turn off the tap when cleaning teeth
WA1 Use a shower or fill the bath half full
EN1 Reduce heating in unoccupied rooms

EN2 Reduce hot water temperature
EN6 Keep heating low to save energy
EN3 Wait until a full load to do washing
EN4 More clothing instead of more heating
EN5 Switch off lights in unoccupied rooms
WA7 Use a sprinkler sparingly, if at all

Recycling
W5 Recycle glass
W6 Recycle newspaper
W7 Recycle cans
W8 Recycle plastic bottles
W2 Donate old household items to charity
W1 Take clothes to charity shops

Behaviour
■ Always
▨ Usually
▧ Sometimes
⬚ Rarely
□ Never

Figure 7.3 Reported behaviour for the Occasional Environmentalists cluster

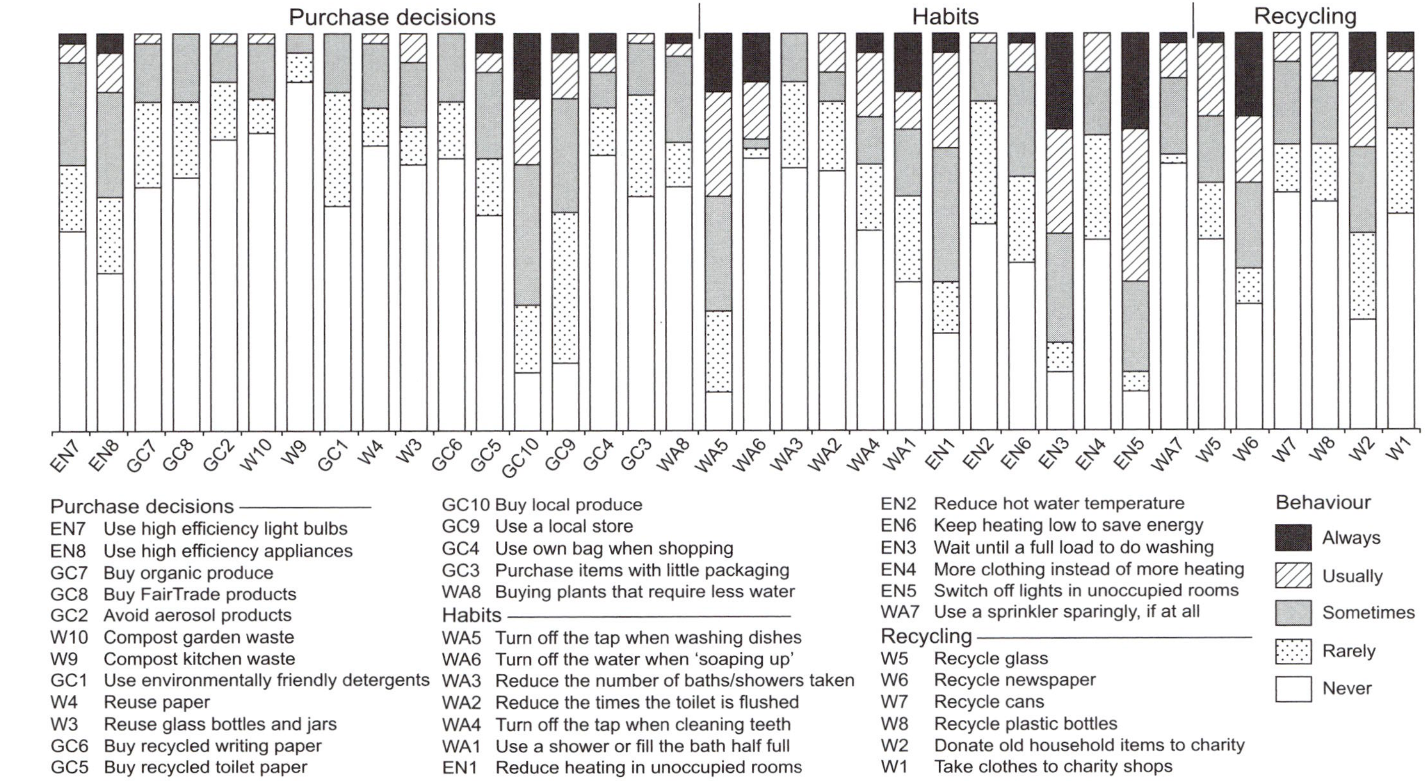

Purchase decisions
Habits
Recycling
EN7 EN8 GC7 GC8 GC2 W10 W9 GC1 W4 W3 GC6 GC5 GC10 GC9 GC4 GC3 WA8 WA5 WA6 WA3 WA2 WA4 WA1 EN1 EN2 EN6 EN3 EN4 EN5 WA7 W5 W6 W7 W8 W2 W1
Purchase decisions
EN7 Use high efficiency light bulbs
EN8 Use high efficiency appliances
GC7 Buy organic produce
GC8 Buy FairTrade products
GC2 Avoid aerosol products
W10 Compost garden waste
W9 Compost kitchen waste
GC1 Use environmentally friendly detergents
W4 Reuse paper
W3 Reuse glass bottles and jars
GC6 Buy recycled writing paper
GC5 Buy recycled toilet paper
GC10 Buy local produce
GC9 Use a local store
GC4 Use own bag when shopping
GC3 Purchase items with little packaging
WA8 Buying plants that require less water
Habits
WA5 Turn off the tap when washing dishes
WA6 Turn off the water when 'soaping up'
WA3 Reduce the number of baths/showers taken
WA2 Reduce the times the toilet is flushed
WA4 Turn off the tap when cleaning teeth
WA1 Use a shower or fill the bath half full
EN1 Reduce heating in unoccupied rooms
EN2 Reduce hot water temperature
EN6 Keep heating low to save energy
EN3 Wait until a full load to do washing
EN4 More clothing instead of more heating
EN5 Switch off lights in unoccupied rooms
WA7 Use a sprinkler sparingly, if at all
Recycling
W5 Recycle glass
W6 Recycle newspaper
W7 Recycle cans
W8 Recycle plastic bottles
W2 Donate old household items to charity
W1 Take clothes to charity shops
Behaviour
Always
Usually
Sometimes
Rarely
Never

Figure 7.4 Reported behaviour for the Non-environmentalists cluster

Mainstream Environmentalists

Cluster 2 (Figure 7.2) was classified as a group of 'Mainstream Environmentalists', partly due to the size of this cluster and also due to the distinctions that were drawn between this group and Committed Environmentalists. Overall, this cluster demonstrated little variation in behavioural commitment compared to the Committed Environmentalists, with the important exception being that under ten per cent stated that they 'always' or 'usually' composted kitchen and garden waste. In general, there were more individuals who never or rarely undertook a range of activities. This was particularly evident with regard to the purchase of fairly traded and organic produce. Nonetheless, the general commitment of this cluster was high, with a wide range of purchase-related activities being undertaken. Accordingly, there was a shift from significant commitment to a more equivocal position, in this case illustrated by the sharp reduction in the number of individuals who composted their waste.

Occasional Environmentalists

Cluster 3 represented a significant shift from the first two groups, being termed 'Occasional Environmentalists' (Figure 7.3). This group were less likely to engage in all of the activities, but especially purchase-related and certain water conservation practices. With regard to purchase decisions, the only items that were undertaken with moderate frequency were waste-related activities such as looking for less packaging, reusing paper and reusing glass. The use of a local food store and the purchase of local foods was also undertaken fairly regularly. However, over 60 per cent of this group stated that they 'never' or 'rarely' purchased fairly traded or organic goods. In terms of habitual activities, water conservation in the home was undertaken with less frequency than for the mainstream group, although habitual actions to save energy were still undertaken frequently, such as waiting until a full load of washing and switching off lights in unused rooms. This cluster also contained larger proportions who either 'never' or 'rarely' recycled waste. The term 'occasional' seems appropriate for this cluster given the greater number of individuals who stated that they sometimes took part in a range of activities, but the notable trend in these data is the shift from the relatively high commitment to recycling to a weaker commitment towards consumptive practices to help the environment.

Non-environmentalists

Finally, Cluster 4 was given the term 'Non-environmentalists' (Figure 7.4). This negative term is framed within the specific context of the reported behaviour measured within this specific study. With this qualification in mind, this cluster stands apart in a number of ways from the previous three clusters. In all but two cases, far fewer than 50 per cent of individuals 'usually' or 'always' undertook each behaviour, with most being under 20 per cent. Those activities where there was more behavioural commitment were related mostly to 'common sense' actions, with overtly pro-environmental behaviours being undertaken with little or no frequency. This group is significant as it demonstrates a distinctly non-committed group of individuals who rarely engage with environmental practices. Once again, there is a clear trend from recycling to purchase decisions. However, even in this case, recycling behaviour is very low, with very few individuals reporting that they 'always' recycled any produce.

Conclusion

These classifications present a new perspective on aggregating individuals into behavioural groups, demonstrating how the study of environmental practices framed within the context of everyday behaviours can yield a series of distinctive trends across alternative groups of individuals. These results present the output from one cluster analysis and therefore represent one *modus operandi* of segmenting this sample from Devon. Given the methodological issues involved in segmentation and the specific nature of this sample, it would be misguided to interpret these results out of context. However, they demonstrate the potential for identifying alternative behavioural trends through the use of segmentation and cluster analysis. In particular, this analysis has revealed the differences that segments reveal between types of environmental practice, with recycling being undertaken most frequently by all the clusters on the one hand, with purchase-related activities witnessing a significant decline from the mainstream group onwards. These behavioural trends present a series of clusters that can be termed 'behavioural groups'. However, the aim of this research was to examine the extent to which these behavioural classifications were also representative of lifestyle groupings. To examine if this was the case, the variables outlined in Figure 4.3 can be examined to discern any differences between the clusters to for each group.

Lifestyle Groups 2: Socio-Demographic Profile

This section provides demographic information regarding the groups identified previously. As noted previously, considerable work on segmentation has focused on the use of socio-demographic measures as a means of both profiling segments and targeting policies for implementing behaviour change strategies. Nonetheless, the links between socio-demographic variables and environmental action are by no means unequivocal as Chapter 5 demonstrated. Accordingly, it must be appreciated that any differences discerned between socio-demographic characteristics and the four clusters are representative of behaviour, rather than necessarily providing the basis for a socially-driven set of policies. Table 7.1 provides demographic data on the whole sample and the four groups of individuals, along with a relevant test statistic showing whether there are significant differences between the four groups. Briefly, the data in Table 7.1 can be summarised as follows:

• The mean *age* of Committed Environmentalists is highest, with the mean age of Non-environmentalists being the lowest. The mean for the whole sample

Table 7.1   Demographic composition of the sample by cluster membership

Variable	Sample	Cluster 1	Cluster 2	Cluster 3	Cluster 4	Test statistic and significance
Cluster label		Committed Environmentalists	Mainstream environmentalists	Occasional environmentalists	Non-environmentalists	
No. in cluster		294	412	505	43	
Age (mean)	49	55	52	46	43	Kruskal-Wallis H = 59.6 (p <0.05)
Gender	Male 35%	Male 35%	Male 31%	Male 38%	Male 50%	Chi-Square = 8.8 (p <0.05)
Tenancy	Owned 74%	Owned 83%	Owned 74%	Owned 71%	Owned 62%	Chi-Square = 23.3 (p <0.05)
	Private Tenant 11%	Private Tenant 5%	Private Tenant 11%	Private Tenant 13%	Private Tenant 19%	
	Local authority 15%	Local authority 12%	Local authority 15%	Local authority 16%	Local authority 19%	
Income	<7.5k 20%	<7.5k 20%	<7.5k 23%	<7.5k 15%	<7.5k 35%	Chi-Square = 29.9 (p <0.05)
(Pounds)	7.5–10k 9%	7.5–10k 20%	7.5–10k 10%	7.5–10k 8%	7.5–10k 6%	
	10–15k 17%	10–15k 11%	10–15k 20%	10–15k 15%	10–15k 9%	
	15–20k 19%	15–20k 15%	15–20k 18%	15–20k 22%	15–20k 12%	
	20–30k 21%	20–30k 19%	20–30k 20%	20–30k 23%	20–30k 20%	
	>30k 14%	>30k 15%	>30k 9%	>30k 17%	>30k 18%	
Education (formal)	None 39%	None 41%	None 41%	None 35%	None 53%	Chi-Square = 21.6 (p <0.05)

Table 7.1 cont'd

Variable	Sample	Cluster 1	Cluster 2	Cluster 3	Cluster 4	Test statistic and significance
Political allegiance	GCSE 27%*	GCSE 20%*	GCSE 30%*	GCSE 29%*	GCSE 19%*	Chi-Square = 43.3 (p <0.05)
	'A' level 17%**	'A' level 18%**	'A' level 15%**	'A' level 18%**	'A' level 16%**	
	Degree 17%	Degree 21%	Degree 14%	Degree 18%	Degree 12%	
	Conservative 16%	Conservative 16%	Conservative 16%	Conservative 15%	Conservative 15%	
	Green 3%	Green 6%	Green 4%	Green 2%	Green 0%	
	Labour 22%	Labour 18%	Labour 22%	Labour 25%	Labour 32%	
	Liberal Democrat 14%	Liberal Democrat 17%	Liberal Democrat 17%	Liberal Democrat 12%	Liberal Democrat 5%	
	Other 2%	Other 2%	Other 1%	Other 2%	Other 2%	
	Would not vote 10%	Would not vote 6%	Would not vote 10%	Would not vote 12%	Would not vote 22%	
	Pass Q 33%	Pass Q 35%	Pass Q 30%	Pass Q 32%	Pass Q 24%	
Membership of a community group	Yes 11%	Yes 17%	Yes 10%	Yes 8%	Yes 8%	Chi-Square = 16.0 (p <0.05)

Notes

* General Certificate of Secondary Education (taken at age 16).

** Advanced Level qualification taken at age 18 or 19.

(49 years old) lies exactly between these two extremes, with mainstream and Occasional Environmentalists having mean ages of 52 and 46, respectively. This gradation demonstrates that those less committed to environmental action tended to be younger, which is statistically significant.

- The overall *gender* balance in the sample is significantly weighted towards females (65 per cent). This should be borne in mind when interpreting the cluster distributions. Nonetheless, there were significantly more males in the non-environmentalist cluster, with the gender balance remains relatively stable in the three remaining clusters. Taking into account the qualification concerning the amount of males and females in the overall sample, it is apparent that the non-environmentalist group contains a far larger number of males than any other cluster.
- Committed and Mainstream Environmentalists tended to have smaller *household sizes* than occasional or Non-environmentalists. A significantly large number of households in these latter groups had more than five individuals in the home, with 17 per cent of Non-environmentalists living in households with five occupants or more.
- *Car access* fluctuated according to the cluster examined and in the overall sample 20 per cent of households reported having no access to a car. No statistically significant differences between car access and cluster membership were found.
- *Home ownership* was high in the overall sample, with 74 per cent stating that they owned their own home, either outright or through a mortgage. Committed Environmentalists tended to own their own home, whilst a greater proportion of Non-environmentalists were either private tenants or rented their home from a local authority: 19 per cent of Non-environmentalists reported living in local authority or housing association accommodation, compared to 12 per cent of Committed Environmentalists.
- The overall sample was well distributed according to *household type*. Committed Environmentalists tended to live in terraced properties, whilst Mainstream Environmentalists were more likely to live in semi-detached homes.
- *Income* also showed a wide distribution across the whole sample, with 20 per cent of the sample earning less than £7,500 a year and 14 per cent earning over £30,000. Non-environmentalists were on significantly lower incomes than any other group, with 35 per cent reporting that they earned less than £7,500 a year, compared to just 20 per cent of Committed Environmentalists with these earnings. However, a significantly higher proportion of Committed Environmentalists earned between £7,500 and £10,000 a year. The higher income brackets were equally spread between groups. The data, whilst statistically significant, demonstrate some interesting trends, not least that Committed Environmentalists were not the most high income-earning group.
- The *formal education* reported by the sample revealed that 385 had no formal qualifications. Committed Environmentalists were less likely to have received any formal education than the average, but equally they were also more likely to have a degree. In the case of Non-environmentalists, a large proportion had received no formal education, with low levels of GCSE, A-level and degree

qualifications. Mainstream and Occasional Environmentalists tended to have GCSE qualifications.

- The *political affiliation* of the sample overall reflected national trends to a point, although the share of Conservative voters was low, potentially accounted for by the large number of respondents who passed the question (33 per cent). Non-environmentalists contained a large amount of Labour voters as well as a significant proportion that did not vote. There were markedly fewer Liberal Democrat voters amongst this group. Committed Environmentalists were more likely to vote Green and Liberal Democrat. They were also the most likely to vote. Mainstream and Occasional Environmentalists represented what one might expect to be the national situation, with Labour the dominant party of choice, followed by the Conservatives and Liberal Democrats.
- Finally, *membership of community, political or environmental organisations* was examined. Committed Environmentalists were significantly more likely to be a member of a community organisation, whilst occasional and Non-environmentalists were least likely to be, with only 8 per cent of Non-environmentalists stating that they belonged to any type of organisation.

Evidently there are more trends than can be discerned Table 7.1, but for the purposes of brevity it is interesting to note that those most committed to a sustainable lifestyle were older, tended to own their home, lived in a terraced property, voted Green/Liberal Democrat and were members of community groups. In contrast, those who were Non-environmentalists tended to be younger, male, on very low incomes, had received less formal education, were less involved in the community and were more likely to be politically apathetic. Nonetheless, the data do reveal that some of the groupings may not be so easily defined by socio-demographic characteristics. For example, Committed Environmentalists were on both high and relatively low incomes. This and other results from this analysis may point to a wider interpretation of environmental lifestyles than can be afforded with a simple socio-demographic analysis. Accordingly, these results are important because they challenge some of the traditional assumptions regarding environmental action, not least that in the case of this sample, those most heavily engaged in environmental action are older and on generally lower incomes. Nonetheless, some of the core assumptions relating to environmentalism are found to exist within this sample, specifically the liberal political attitudes of environmentalists and the gender imbalance of pro-environmental individuals. Most importantly, the data provide general support to the notion that the trends described were graduated from the Committed Environmentalists to the Non-environmentalists, with this last group standing out the most. This will now be examined in relation to the remaining variables in the conceptual framework of environmental behaviour (Figure 4.3).

Lifestyle Groups 3: Social and Environmental Values

The literature review in Chapter 5 demonstrated the somewhat equivocal links between social values and environmental behaviour. Stern et al. (1995) and Corraliza and Berenguer (2000) have claimed that using Schwartz's (1992) inventory of value constructs according to his 'pro-self' – pro-social' and 'open to change – conservative' value continuums can distinguish pro-environmental individuals as being more pro-social and open to change. This research measured a series of social value constructs (Schwartz, 1992; Table 6.3). To examine the nature of the different constructs and to provide the basis for constructing four value types allied to Schwartz's (1992) categorisation, the social value measures were placed into a principal components factor analysis to examine the empirical links between the different constructs. As reported in Chapter 6, the goal of factor analysis is both to reduce the number of variables and also to examine conceptual links between these items.

Table 7.2 demonstrates that four factors emerged related to Schwartz's original conception of social values. These have been classified according to altruism (pro-social), egoism (pro-self), openness to change and conservative. The value constructs in each factor provide good evidence to suggest that Schwartz's (1992) conceptualisation of value continuums is a useful means by which to categories his values. Accordingly, the items for each social value factor were summed to make four new value scales. To investigate whether these differed according to environmental behaviour, the differences between each behavioural cluster and each social value factor were investigated statistically using the Mood Median test. These are shown in the last column in Table 7.2 and have an associated p-value and mean score.

These data indicate some key trends with regard to social values. In particular, there would appear to be a clear divide between Non-environmentalists (Cluster 4) and the remaining sample. Those least committed to environmental action were more likely to value wealth, social power and be less committed to unity. These data do not provide evidence that there is a clear 'altruistic – egoistic' continuum of values on which the environmentalist can neatly be placed. Rather, given the nature of the questions posed and the social desirability of answering questions in certain ways, a skewed pattern has emerged that only reflects the (significant) divergence between Non-environmentalists and those who are relatively committed. In summary, social values do not generally appear to be related to environmentalism, but this relationship is skewed and may need further investigation.

The research also used an adapted version of Dunlap and Van Liere's (1978) New Environmental Paradigm scale to examine the operational and relational aspects of environmental values, in an attempt to examine the role of such values in predicting commitments to environmental action. The environmental values measured in the research were also placed into a principal components factor analysis, which sought to understand the relationships between the items that measured value constructs. Table 7.3 provides the results from these analyses. It can be seen that two main factors emerge, one of which contained items that relate to a biocentric–ecocentric perspective and the other to an anthropocentric–

Table 7.2   Factorially-defined social values of the sample by cluster membership

Factor	Variables included	Means				Variance (eigenvalue)	Per cent variance	Mood median
		Cluster 1	Cluster 2	Cluster 3	Cluster 4			
Altruistic	Loyalty Honouring parents Equality Social justice Enjoying life Helpful	26.3	26.5	26.4	26.6	3.1	20%	5.8 (p > 0.05)
Openness to change	Varied life Exciting life Curious	11.5	11.3	11.6	11.6	2.5	16%	2.6 (p > 0.05)
Conservative	Social order Obedience Authority Unity	15.5	15.5	15.3	14.9	2.1	13%	7.8 (p < 0.05)
Egoism	Wealth Social power Influential	8.7	9.0	9.1	9.7	1.8	11%	9.8 (p < 0.05)
Total variance							60%	

Notes

Individuals were asked how important each value was to their own life, rating each from 1 (very unimportant) to 5 (very important).
Each factor mean is a composite score of the items in the factor and divided by the number in the sample.

Table 7.3 Factorially-defined environmental values in the sample by cluster membership

Factor	Variables included	Means				Variance (eigenvalue)	Per cent variance	Mood median
		Cluster 1	Cluster 2	Cluster 3	Cluster 4			
Faith in growth: anthropocentrism	There are no limits to growth for nations like the UK Modifying the environment seldom causes serious problems Science will help us to live without conservation Humans were created to rule over nature	15.4	14.7	14.8	13.3	2.3	24%	12.6 ($p < 0.05$)
Spaceship Earth: biospherism	The balance of nature is delicate and easily upset The Earth is like a space ship, with limited room and resources Plants and animals do not exist primarily for human use One of the most important reasons for conservation is to preserve wild areas	16..6	16.3	16.1	15.1	2.2	22%	10.2 ($p < 0.05$)
Ecocentrism-technocentrism	Technology will solve many environmental problems Exploitation of resources should be stopped	6.8	6.9	6.8	6.2	1.2	12%	3.9 ($p > 0.05$)
Total variance							58%	

Notes

Individuals were asked to rate their agreement with each statement from 1 (strongly disagree) to 5 (strongly agree).
Each factor mean is a composite score of the items in the factor and divided by the number in the sample.

technocentric viewpoint. The third factor contains operational values only. These factors were summed to form scales in the same way as the social value factors.

With regard to environmental values, a far clearer and uniform pattern emerges in terms of the differences between the scores on these factors and the behavioural clusters. The last column of Table 7.3 presents the results of statistical tests to examine the differences between each factor and cluster. The mean values show a clear trend in the data, with higher scores for those in the more committed clusters on the environmental value scales. Although the final factor (Ecocentrism-Technocentrism) is more equivocal, the trends in the first two factors are much clearer.

To a certain extent, social values differ according to behavioural category, with Committed Environmentalists more likely to hold values that encapsulate unity with less emphasis on personal wealth. However, environmental values demonstrate a more definitive distinction between the behavioural groups, especially the Committed Environmentalists. They believed in the intrinsic value of nature and in limits to growth. Accordingly, the previous research linking environmental values to behaviour is supported, giving credence to the argument that those not involved in environmental action tend to believe that humans have greater value than nature and that technology provides a means by which to resolve environmental dilemmas. Such a finding enables social researchers to appreciate the effect of embedded values in everyday lifestyle choices and may assist in the development of further lifestyle adjustment programmes. We will now turn to examining the final set of variables in Figure 4.3: psychological variables.

Lifestyle Groups 4: Psychological Variables

The psychological variables were measured in the questionnaire using a series of Likert agreement statements, which provided too large a data set to deal with in raw form. It was therefore decided that, as for the social and environmental values, the psychological items would be placed into a principal components factor analysis. Unlike the social and environmental value factors, there was a greater emphasis on the exploratory nature of the factor analysis, without any preconceived ideas about how this large number of items would load onto alternative factors.

Table 7.4 provides the results of these factor analyses. Fifteen factors were identified. The second column in Table 7.4 provides a summary of the composite variables that were present in each factor, which in turn determined the name of the factors provided in the first column of the table. The four columns of mean values present the average score on each factor (comprising the summed scores of all items in each factor) and is given for each cluster. Once again, a statistical test determining any significant differences between the four clusters is provided in the final column. It is worth noting that many statements in the questionnaire were posed in an 'anti-environmental' direction, acting as 'check' statements for each construct. Table 7.4 presents means based all a pro-environmental direction, where statements have been recoded accordingly before analysis.

Table 7.4   Factorially-defined psychological variables in the sample by cluster membership

Factor	Salient composite variables (expressed in pro-environmental direction)	Means				Mood median
		Cluster 1	Cluster 2	Cluster 3	Cluster 4	
Concern and commitment	Environmental concern Time to help the environment Self efficacy in helping the environment Personal responsibility for environmental problems	27.7	26.5	25.1	22.0	69.6 ($p < 0.05$)
Moral motives	Moral obligation to help the environment Social acceptance of environmental behaviour Self presentation of environmental action a positive quality	18.7	18.6	17.2	15.2	62.3 ($p < 0.05$)
Environmental beliefs	Beliefs in efficacy of actions	17.0	16.7	16.0	14.9	45.1 ($p < 0.05$)
Influence of price	Willingness to pay more for environmental products	11.3	10.8	10.0	9.5	32.5 ($p < 0.05$)
Satisfaction in helping the environment	Various satisfactions from environmental action	18.9	19.1	18.6	18.7	9.4 ($p < 0.05$)
Logistics and convenience	Convenience of helping the environment	10.6	10.0	8.9	7.8	78.1 ($p < 0.05$)
Green consumer beliefs	Importance of health, safety and local issues to helping the environment	15.8	15.3	14.4	13.4	84.9 ($p < 0.05$)
Comfort	Willingness to sacrifice home comforts to help the environment	7.0	6.9	6.4	5.4	42.9 ($p < 0.05$)
Environmental rights	Belief in strong environmental rights	10.2	10.0	9.3	8.1	37.0 ($p < 0.05$)

Table 7.4 cont'd

Factor	Salient composite variables (expressed in pro-environmental direction)	Means				Mood median
		Cluster 1	Cluster 2	Cluster 3	Cluster 4	
Awareness of norm to help the environment	Awareness of the normality of environmental actions	6.2	6.3	5.8	5.1	23.5 ($p < 0.05$)
Trust and responsibility	Trust in environmental information Importance of personal responsibility	9.0	9.5	9.0	9.1	12.3 ($p < 0.05$)
Extrinsic motivation	Rejection of extrinsic motives to help the environment	6.8	6.6	6.4	6.4	10.7 ($p < 0.05$)
Personal instinct	Trust in one's own instinct regarding the environment	3.8	3.7	3.6	3.9	12.4 ($p < 0.05$)
Brand loyalty	Willingness to change brand labels to help the environment	2.6	2.7	2.7	3.0	3.5 ($p > 0.05$)
Personal threat	Threat of environmental problems to the self	3.9	3.8	3.7	3.5	17.8 ($p < 0.05$)

Note

Each factor mean is a composite score of the items in the factor and divided by the number in the sample.

The Psychology of Environmental Practice

An examination of Table 7.4 reveals some interesting points in their own right relating to the construction of psychological factors and environmental action:

- The first factor (*Concern and commitment*) involves a range of items relating to both levels of environmental concern, but also measures of self-efficacy, in particular the time that individuals felt they could commit to helping the environment and the personal responsibility they held for undertaking environmental behaviours. This factor appears to be quite broad in nature, but it touches on a clear notion of strong environmental concerns (both an overt statement of concern and also a reflection of this in terms of time commitment). The relative importance of 'time' to this factor also relates to the commitment individuals are willing to provide in relation to the personal responsibilities individuals are willing to acknowledge. Overall this factor represents items in the questionnaire that reflect specific commitments to the environment.
- The second factor (*Moral motives*) is clearly a measure of the normative acceptance and 'social pressure' that individuals may feel to participate in certain environmental behaviours. To a certain extent, a moral obligation to help the environment relates to the personal responsibilities that individuals feel towards environmental issues. However, in this instance, it appears that morality is connected to both the social acceptance and the desirability of environmental practices. Accordingly, moral obligations to help the environment may be related to, or be a product of, the increasing social desirability of undertaking environmental behaviours. It is interesting to note that factorially, this construct is differentiated from an awareness of the norm to help the environment (see below), which may indicate that the 'Moral motives' factor is representative of a wider social construct recognising the overall desirability and social acceptance of environmental practices, rather than the influence of specific individuals.
- The *Environmental beliefs* factor was comprised of four statements in the questionnaire that all posed a series of scenarios relating to the potential outcomes of environmental behaviour, such as "saving energy helps reduce global warming". This is a factorially distinctive construct highlighting the links between the diverse, but nonetheless related psychological importance of believing in specific outcomes of a behaviour.
- *The Influence of price* was another distinct construct, acknowledging the importance to individuals of a series of items that posed questions both concerning the role of price in guiding consumption choices and also the potential trade-offs between price and environmental priorities that individuals were willing to make.
- The *Satisfaction in helping the environment* factor measured a range of satisfactions that could potentially be derived from helping the environment, such as 'It makes me feel good when I do something to help the environment'. This factor taps into a number of motivational factors that provide individuals

with a range of intrinsic motivations to help the environment, as opposed to extrinsic (e.g. financial) incentives.

- *Logistics and convenience* has been highlighted by a large number of researchers as being a major factor that influences uptake of environmental action. In this research, three items from the questionnaire comprised a 'convenience' factor, indicating that the overall convenience of helping the environment, along with more specific concerns such as storage space for recyclable items, was a distinct construct.

- As noted in the literature review in Chapter 5, specific attitudes towards the consumption of green products have been identified by a range of researchers (e.g. Roberts, 1996). Table 7.4 demonstrates that in this research, a series of attitudinal constructs such as the importance of health, safety and locally-defined issues were all closely related terms *Green consumer beliefs*. This indicates that attitudinal responses to key questions relating to green consumption are indeed closely related and comprise a genuinely 'green consumer' factor.

- Related more closely to energy and water conservation, the *Willingness to sacrifice comfort* factor comprised two items highlighting the importance of a willingness to sacrifice some comforts and conveniences to help the environment.

- *Environmental rights* was a factor comprised of statements that related specifically to the right individuals have to access certain environmental resources and the extent to which this should be limited by the (overriding) concerns for environmental protection.

- In contrast to the moral motives factor, the *Awareness of the norm to help the environment* factor comprised two items that highlighted the importance of both the awareness of 'significant others'' behaviour and the acceptance of their behaviour as 'normal'. This highlights the key difference between a wider social and moral construct, emphasising the overall social desirability of environmental action and the more specific influence of neighbours and friends influencing attitudes towards the environment. Perhaps the most effective example of the influence that awareness of others' activity has relates to kerbside recycling, where neighbours can 'monitor' the activity of fellow residents and provide social pressure to recycle.

- The *Trust and responsibility* factor comprised items that joined two interesting constructs. First, the trust individuals have in official sources of information and second, the personal responsibility that individuals feel to help the environment. The connectivity between these two items presents the potential to conceptually link the trust that individuals have in information provided by official sources, such as the government and the extent to which this impacts on the power of personal responses to help the environment. Given that this factor comprised three items, the means are quite low, indicating that those with low levels of trust also had similarly low levels of personal responsibility.

- In contrast to the intrinsic (satisfaction) motivation factor detailed above, the *Extrinsic motivation* factor emphasised the need to use incentives and other external measures to encourage individuals to help the environment.

- Finally, three items formed factors in their own right, relating to the importance of *Personal instinct* regarding environmental issues, rather than official narratives of environmental issues, as well as 'Brand loyalty', which was a consumption item focused around the willingness to change brand if there was an environmental alternative. Finally, the 'Personal threat' factor measured the importance that threats to both oneself and the family has in motivating environmental behaviour.

These factors provide some excellent comparisons to the literature review presented in Chapter 5. In many cases, the constructs identified as psychological factors in the literature are corroborated in the results from Table 7.4. However, the use of statements which relate to environmental attitudes more generally, rather than relating these to specific environmental actions has presented some interesting contrasts to previous research. In the first instance, this research has identified a difference between the moral motives to help the environment and the specific normative influences acting on individuals. Researchers such as Chan (1998; 2001) have emphasised the role of social pressure, but by incorporating notions of social desirability (Sadalla and Krull 1995) this research has demonstrated how the over-arching social context of environmental action can be seen as an alternative psychological factor to the specific social norms which individuals are subject to in specific situations, such as recycling behaviour.

A second major contrast to previous research is the split that appears evident between alternative types of self-efficacy. Oskamp et al. (1991) among others has highlighted the importance of this construct, embedded in the Theory of Planned Behaviour (Ajzen 1991). The results from Table 7.4 suggest that two alternative constructs may be appropriate. First, a factor that combines environmental concerns (as conventionally defined by Weigel and Weigel 1978) and overall commitment towards helping the environment. This is in contrast to a more specific set of self-efficacy factor represented by specific 'convenience' issues. Once again, a distinction can be drawn between the general expressions of environmental concerns and commitments and more specific attitudes towards acting out a particular activity.

A third and final distinction between previous research and this work is the incorporation of green consumption items into a wider study of environmental action. The work of Roberts (1996) was instructive in compiling the green consumption items for the questionnaire and it is interesting to note that these consumption variables have mostly remained as distinct factors, rather than being incorporated into the wider set of psychological factors. This supports the hypothesis presented earlier in this chapter that environmental practices are indeed focused around consumptive, habitual and recycling practices and that attitudes are representative of these practices.

Accordingly, the factor analysis of psychological items in the questionnaire yielded a series of distinct and factorially valid constructs that enable an evaluation of the conceptual framework presented in Figure 4.3. Most importantly, Figure 4.3 appears to play down the complexities of psychological factors and there is clearly a need for researchers to recognise the realities of psychological dimensions of environmental action. These dimensions are multilayered and related, being both behaviour-specific and socially contextual in nature. They also emphasise the need to examine the relatedness of constructs such as trust and responsibility, both key factors that can determine the uptake of environmental action.

Segmenting the Psychology of Environmental Practice

The question to emanate from these data relates to the potential for distinguishing between different types of behavioural grouping according to psychological factors. In essence, is it possible to identify 'lifestyle groups' from behavioural groups, on the basis of these psychological attitudes?

With the exception of one factor (Brand loyalty), there was a statistically significant difference between the behavioural groupings for each factor. Given that the scales were coded in order to show a pro-environmental attitude in all cases, higher scores (and therefore means) represent a more pro-environmental position. Although means are not the most effective way of showing such information (although the data are on ratio scales, they are constructed from Likert data which do not have equal distances between data points), the comparisons may be more evident than showing 15 bar charts. The data are conclusive in all cases, with the highest scores being achieved for the Committed Environmentalists and the lowest for the Non-environmentalists. As noted earlier, the apparent fluctuation in mean values between each factor reflects the number of items in each factor, rather than the strength of agreement and therefore pro-environmental attitude.

The greatest differences, indicated by higher mood median statistics (all $p < 0.05$), indicate that concern and commitment, moral motives, logistical factors and green consumer attitudes provide scope for differentiating between different types of environmentalist. To this extent, these factors may provide the basis for a differentiation between the four clusters. Four key themes emerge from these differences. First, high levels of concern, which are illustrated by the personal responsibility and moral obligation felt by respondents to help the environment within the 'Concern and commitment' factor illustrate that personalisation of an environmental problem and a genuine moral obligation to act is a motivator for environmental action. This finding contributes to a number of debates highlighted in the literature (see, for example Derksen and Gartell 1993 and Guagnano et al. 1995) that have contested the role of environmental concern in determining levels of environmental action. This research would support the assertion that environmental concern can be used to distinguish between the commitment of individuals who help the environment.

Second, a further theme to emerge from the analysis of means in Table 7.4 relates to logistical factors and the ease with which individuals felt they could be pro-environmental. A belief that helping the environment takes less effort and is worthy of a time commitment is evidently significant. This supports the notion that individuals who display greater levels of environmental action perceive fewer barriers to that behaviour and will more actively seek to overcome any potential conflicts accordingly.

Third, there is the importance highlighted by the mean scores on the 'Moral motives' factor that demonstrate the significance of a perceived social acceptance of a behaviour, enabling individuals to take part with minimal personal sacrifice to their self-presentation (Sadalla and Krull 1995). This supports the notion that less pro-environmental individuals perceive a certain social stigma about participating, reflecting an underlying social discourse which places environmentalism outside the mainstream of socially desirable activities. Evidently, the measures used in the questionnaire to assess social desirability (see Chapter 6) were crude and more research is required in this area. However, it does appear that Committed Environmentalists were either tapped into a social network that highlighted the importance of acting in an environmentally-friendly manner or they were able to overcome any social stigmas relating to their environmental practices.

Finally, the fourth theme to emerge relates to the specific importance of green consumer beliefs, with those more concerned with health and safety issues and the importance of local produce being more pro-environmental. This highlights the importance of specific consumption issues in differentiating between Committed Environmentalists and the other segments.

Looking more broadly at the factors in Table 7.4, a further trend in the data relates to the significant, if less defined, difference between factors pertaining to outcome beliefs, price, comfort and environmental rights. This relates to two further issues. In the first instance, the efficacy of a given activity is evidently highest amongst those who were Committed Environmentalists, indicating that being confident in the ability of their personal actions to effect a tangible environmental outcome is significant in motivating behaviour. A second theme reflects the wider debate surrounding personal sacrifices on the one hand and willingness to pay for environmentally-friendly products on the other. Those who tended to be less willing to pay for environmentally-friendly products were less inclined to help the environment, a theme that is demonstrated with regard to the willingness of individuals to sacrifice home comforts. It is also interesting to note that those who were least willing to pay more for environmentally-friendly products were also those who were least concerned about health, safety and local issues when purchasing goods.

Some weaker differences evident in the data, pertain to the trustworthiness of environmental information, satisfactions and brand loyalties. The data suggest that there are differences between the behavioural groups according to the trustworthiness of information, but that these do not reflect behavioural commitment in the more linear fashion shown above. Accordingly, Committed Environmentalists may be less inclined to trust information provided than Non-environmentalists. Indeed, significant differences were reflected between the groups and scores on the satisfactions factor, even though there is little discernible difference between committed and Non-environmentalists. Finally, there was no significant difference between the willingness to change brand loyalty and behavioural commitment.

These data suggest that the dominant themes in the literature relating to attitudes towards environmental action are generally supported. There are strong differences in behavioural commitment according to environmental concern, social acceptance, convenience and green consumer motivations. However, the influence of satisfactions and trust in environmental information is questionable. Overall, at first glance the 'environmentalist' is a highly concerned individual, motivated by a range of issues, who is confident in the outcome of their actions and finds helping the environment relatively simple and socially desirable.

From Behavioural Groups to Lifestyle Groups?

The analysis of each segment's properties in terms of socio-demographic profile, social and environmental values and psychological variables presents a snapshot of each behaviourally-defined group. However, as stated in the objectives of the research in Chapters 5 and 6, the overriding empirical and conceptual goal was to define lifestyle groups, on the basis of the wide range of variables in Figure 4.3. The analysis of data provided in this chapter would suggest that, although the clusters of environmentalists were defined by behavioural commitment to the 36 activities in the questionnaire, the differences observed in terms of key lifestyle variables would support the notion that the research has identified key lifestyle groupings.

The formation of lifestyle groups necessitates a shift in emphasis from defining groups purely in terms of their reported actions towards an appreciation of the lifestyles led by these distinctive segments. This involves a greater emphasis on a holistic understanding of both lifestyle practices but also values and attitudes. In summary, therefore, the following sections provide an overview of the four 'lifestyle' groups identified in the chapter.

The Committed Environmentalist

The committed environmentalist is a highly motivated, interested and committed individual who has a key stake in environmental issues. They display a wide range of behavioural commitments designed to reduce environmental impacts across a range of topics, from waste management to ethical trade. They are 'overtly' environmental, defining their lifestyle in many ways in terms of their experiences and practices. They hold pro-social values and are open to new ideas, innovations and changes that increase the sustainability of their behaviour. They also hold pro-environmental attitudes. Overall, this group believes that helping the environment is socially desirable and that there are few barriers to helping the environment. They look for environmentally-friendly goods and are willing to pay more for environmentally sustainable produce. Demographically, this group comprises both wealthy individuals, as well as people on relatively low incomes, but overall this group tends to be older, with a high level of formal education and displaying political activity, with affiliations towards the Green Party and Liberal Democrats.

The Mainstream Environmentalist

The mainstream environmentalist displays a similar behavioural commitment to the committed environmentalist, with the notable exception of composting behaviour. This distinction is significant because it reflects the generally lower level of commitment illustrated across both values and attitudes. The motivations to help the environment may be different for this group, given their lukewarm support for paying more to help the environment and the decline in support for strongly ecocentric values. Accordingly, this group may not define their lifestyle according to their environmental action, but they do act out and share many of the positive attitudes of the Committed Environmentalists. Demographically, this group is younger than the committed group, with lower levels of formal education and political activity. Overall, this group is a mainstream segment of society, representing many individuals who are sympathetic to environmental issues and do 'what they can' to help the environment.

The Occasional Environmentalist

The third cluster (Occasional Environmentalists) is a much less committed group behaviourally, with very few green purchase-related activities undertaken with regularity and a focus on 'easy' energy and water saving habitual actions in the home. They are generally recyclers. This group places a greater focus on pro-self values and is minded to support anthropocentric–technocentric views concerning the environment. In general, they have more negative attitudes towards helping the environment, believing that there are more social 'risks' in undertaking environmental action, along with a series of barriers to environmental action. Overall, this group is a set of pragmatists; in other words, they are willing to undertake environmental action in cases where it requires little effort and does not jeopardise their existing lifestyle.

The Non-environmentalist

The final grouping are the most distinctive. They rarely undertake environmental action and this is mirrored by their generally negative environmental attitudes. They perceive helping the environment to involve a series of social risks which make environmental action socially undesirable. They are unwilling to pay more for environmentally-friendly products or sacrifice comfort to save energy or water. Indeed, they are the most sceptical about the impact of helping the environment. Despite these generally negative aspects, they do report deriving satisfaction when they do help the environment, which indicates that outside of a social situation, these individuals may be willing to undertake environmental action. Demographically this group are the youngest, with a higher proportion of males, on very low incomes, with little formal education, living in privately rented or local authority housing and displaying a large amount of political apathy. Accordingly, this group is perhaps the most challenging from the perspective of behaviour change. Although they are a small number within this sample, they may have been the least likely to answer the survey and could therefore represent a wider group in society. Indeed, from one perspective, it could be argued that the consumption and behavioural activities of this group might be more damaging to the environment overall. Accordingly, this lifestyle group represents a 'hard core' which requires more study. Nonetheless, it would be misleading to simply classify a whole social group as 'Non-environmentalists'. It must be remembered that a limited number of variables were examined in this research and, whilst these individuals may present a group that *could* be doing more damage to the environment by their inaction, they could equally be, by virtue of their social situation, low consumers of energy, water and products that have, for example a high number of associated food miles. To this end, this group may represent a lifestyle segment that do not consume energy and water and produce waste at the same rate as more affluent groups.

Forming lifestyle groups in this way can be extremely useful in terms of plotting the values, attitudes, behaviours and social characteristics of individuals onto a type of 'continuum'. Indeed, the data presented in this chapter would indicate that such a continuum does have validity. Nonetheless, there are limitations of this approach. First, although we have examined lifestyle characteristics for each segment, the actual groups are based on behavioural traits. Accordingly, if the analysis were undertaken by placing all of the other variables from Figure 4.3 into the cluster model, a different set of results may have been achieved. However, this study was focused around examining behaviour and therefore the implication that lifestyle groups have been formed on the basis of behavioural segments is partly unsupported, but the differences shown in this chapter do imply that behavioural differences may be representative of lifestyle differences.

A further limitation to this type of segmentation is that cluster models are normally based on studies that utilise data from one point in time. Thus, the research on which the analysis in this chapter is based was undertaken during 2002. There is little doubt that behavioural levels and, potentially, lifestyle segments will have shifted. Indeed, there is a need to develop longitudinal research that can examine the impact of lifecycles on environmental action and the extent to which individuals move from one lifestyle group to another. This is potentially the case with the data in this research, where those most committed were generally older. It is possible that those in the non-environmentalist group might 'move' into a different lifestyle group at a different stage in the life cycle. However, although this sounds compelling, the social composition of the groups does indicate that these are fairly fixed social and lifestyle segments.

Finally, the major limitation of segmentation, as will be seen in Chapter 9, is the ability to transfer what are interesting findings into potentially useful and usable policy instruments. As can be seen from the data presented in this chapter, targeting individuals with these lifestyle groups would, at first glance, be problematic given the social groups that they represent. Accordingly, new and innovative ways have to be examined to provide the basis for effectively using segmentation, as has been operationalised by the commercial sector for decades, in the promotion of environmental action.

Despite these limitations, segmentation provides a means by which to examine and understand the complexities of environmental action within the context of everyday practices and lifestyles. This chapter has demonstrated the value of examining a range of variables to identify the 'lifestyles' of individuals who do and do not help the environment. Yet to formulate policies for behaviour change, the specific antecedents of each environmental practice for every lifestyle group needs to be appreciated. This is the topic on which we will now focus.

Chapter 8

The Value-Action Gap

Introduction: Intentions and Behaviour

The publication of the 2005 Sustainable Development Strategy (DEFRA 2005) highlighted the realisation in central government that changing behaviour was not merely a question of providing information and expecting a commensurate behavioural response. Chapter 3 of the Strategy began by outlining a series of research projects that demonstrated the complexities of behaviour change. Central to the emerging discourse of 'complexity' was a recognition that the linear model of policy-making was problematic. The so-called A-I-D-A (Awareness–Information–Decision–Action) model of behaviour change had been highlighted by a number of government-sponsored research projects (e.g. Darnton 2004a; 2004b; Cabinet Office 2004). In Chapter 4 we examined critiques of this rationalistic approach provided by geographers. Indeed, we examined how a social-psychological examination of this 'problem' could be used effectively to examine the underlying influences on behaviour and the potential for behaviour change.

Central to a shift away from the A-I-D-A model of behaviour has been the focus on what motivates behaviour and what barriers serve to prevent certain activities (DEFRA 2005). This refocusing has been of great utility in expanding the debate (see for example the raft of DEFRA research on behaviour change, reported in Barr et al. 2006b). However, research has often filled the vacuum created by a critique of the linear model of behaviour change with what can be termed 'ad hoc' studies of behaviour change, which are not framed within a particular theoretical context. This can be problematic, not least because the ability to compare and incrementally move the debate forward can be impaired. Indeed, they rarely attempt to recognise the importance of the 'value-action' gap, which has been so central to both a re-examination of policy and detailed academic discussion. Ignoring the discord between intentions and action and simply focusing of 'what influences behaviour' can be misleading, not least because previous research (Ajzen 1991) has demonstrated that the gap between intentions and actions is complex. In some cases, certain behaviours will have a very small gap between stated intention and action, whilst others may engender a positive intention from respondents, not a weak behavioural commitment. In addition, where a strong link exists between intentions and action, there may be certain crucial factors that act to transform intentions into behaviours and other variables that create positive intentions without directly influencing behaviour. Understanding all of these relationships is vital to an understanding of environmental action, as highlighted in Figure 4.3.

As noted in Chapter 4, within the social-psychological literature the use of Fishbein and Ajzen's (1975) Theory of Reasoned Action (TRA) has been prolific and Ajzen's (1991) more recent Theory of Planned Behaviour (TPB) has been applied in a range of research contexts. The TPB's emphasis on the critical relationship between intentions and behaviour provides the central element of the conceptual framework used in this research (Figure 4.3). Accordingly, this chapter will examine the efficacy of using the framework to answer the third empirical objective of the research, which related to examining the factors that motivate and act as barriers to behaviour change (Figure 6.8). This will be undertaken in three parts. First, the chapter will outline the basis for the analytical techniques utilised to derive the significant factors influencing behaviour, notably multiple regression and path analyses. Emphasis will be placed on understanding the key factors that serve to drive behaviour and intention using a series of Stepwise regression techniques, which can be used to identify significant variables in a model. These can then be plotted as 'path diagrams' to illustrate the strength of relationships present.

The chapter will then proceed to examine two key sets of results that serve to illustrate both the value of these analyses in understanding behaviour, but also the critical importance of examining behavioural segments. Accordingly, the second part of the chapter will present the results of three path analyses, which examined the factors that influenced the three types of behavioural practices we identified in Chapter 6. These analyses will demonstrate the value of classifying behaviour into the three groups of purchase decisions, habits and recycling. However, in the third part of the chapter the value of examining lifestyle groups (Chapter 7) will also be outlined. On the basis of both the behavioural classifications and the lifestyle groups identified previously, path analyses will demonstrate the utility of segmenting the population and illustrate the very different factors that influence behaviour for alternative groups. Once again, as emphasised in Chapter 7, the analyses presented in this chapter are an illustration of the value that can be derived from examining environmental practices and lifestyles. The reader should bear in mind that the factors identified as significant only relate to the sample survey on which this research is based. Accordingly, an alternative sampling strategy and framework could derive different results.

Analysing the 'Value-Action' Gap

The questionnaire on which this research was based collected data relating to a large number of variables and these were all measured using quantitative scales. The conventional means by which such data has been analysed within an explanatory context has been through the use of multiple regression analysis. The aim of regression analysis is to explain the variation in a dependent variable. In other words, regression seeks to understand how a range of independent (or 'influencing') variables can be used to predict the variation and hence the value of a dependent variable. In the case of this research, the dependent variable is behaviour and the independent variables are the other factors outlined in Figure 4.3. However, as the reader will note, although behavioural intention in the framework is an independent variable, predicting behaviour, it is also a dependent variable, subject to the influences of social and environmental values, situational variables and psychological factors. The issue of how to deal with this apparent problem will be outlined below.

As Wheeler et al. (2004) note, simple regression involves using a regression equation to estimate the value of a dependent variable from the value of an independent variable. Given that the two variables are not the same and therefore not identical, the level of explanation provided by the independent variable is usually expressed as a percentage. Accordingly, we might state that intentions explain 56 per cent of reported behaviour. Within the social sciences, researchers have worked according to a series of conventions relating to the levels of explanation offered by regression models, with an explanation of less than 25 per cent considered to be weak, 26 per cent to 50 per cent moderate, 51 per cent to 75 per cent good and over 75 per cent excellent. Given the complexity of behaviour, we would not anticipate that models would reach levels of explanation over 75 per cent.

The complexity of behaviour necessitates that multiple regression is used to examine the basis for explaining reported behaviour. By using more than one independent variable in a regression analysis, the level of explanation should be increased overall. However, where a large number of variables might be used to predict a construct as complex as behaviour, it is likely that a large number of variables within a model will be insignificant to its overall prediction. Accordingly, simply inserting a large number of variables into a regression model can lead to misguided conclusions if the relative importance of such variables is not accounted for. A number of analytical techniques exist for overcoming this problem, depending on the theoretical assumptions being made concerning the nature of the work being undertaken.

Some researchers (e.g. Oskamp et al. 1991) have utilised hierarchical multiple regression techniques, which are based on a prescribed notion of the significance of particular factors in a model. For example, researchers attempting to predict behaviour may specify that behavioural intention is maintained within a model irrespective of its significance, whilst adding and deleting other variables related to their statistical significance. This is normally undertaken when researchers are convinced of the efficacy of a specific variable. More common are what can be termed 'setwise' or 'stepwise' regression models. The first technique seeks to add and remove a complete combination of independent variables in stages to achieve the highest explanation (in per cent) and also the highest significance of the model (expressed as an *F* ratio). Setwise regression is a very useful technique, therefore, for examining the influence of independent variables on a dependent variable. However, it can normally only be used where the number of independent variables is under 30 in number. Accordingly, for this research, 'stepwise' regression was utilised. This technique uses the same process of adding in and removing independent variables in 'steps' to achieve the highest level of explanation.

Accordingly, the results from a stepwise regression analysis are presented as a series of stages (or steps) that have a different number of independent variables progressively showing higher explanation levels. The end product is a reduced set of independent variables each providing a significant contribution to the model.

Each model therefore has an associated explanation termed 'R-squared-adjusted', representing the percentage explanation the model offers. An associated *F*-value provides the overall significance of the model, representing the ratio of explained variance to unexplained variance. This has an associated p-value, with a significance level set at 0.05. For each independent variable that has been identified as significant to the prediction of the dependent variable, a regression coefficient is also provided. This represents the unit movement in the independent variable for a one-unit move in the dependent variable. Using such 'unstandardised' regression coefficients provides difficulties when analysing data measured on different scales (for example age in years vs. agreement using a five-point scale) because comparing between these measures to determine the importance of each variable requires them to be measured on the same scale. What most researchers do to overcome this problem is to standardise their data before analysis, which ensures that regression coefficients presented standardised (or Beta weights) results, which can be directly compared. In addition to the regression coefficient, each significant independent variable has an associated t-value, which is a measure of whether the variable is indeed a significant contributor to the model, as well as an associated p-value.

Accordingly, stepwise regression can be used to identify significant contributors to the prediction of a given dependant variable. However, as was noted previously, the framework in Figure 4.3 would indicate that two multiple regressions would be required to successfully analyse the data. First, a model would be required to examine the influence of social and environmental values, situational characteristics and psychological factors on behavioural intention. Second, a model would be needed to examine all of these influences, plus behavioural intention, in predicting behaviour. Such an analysis cannot be undertaken in one sequence, but through the use of a technique called 'path analysis', the two models can be linked.

Bryman and Cramer (2006) have outlined the essential elements of path analysis, which involves plotting the results of a number of regression analyses onto a diagram, using arrows to represent the strength of relationships between variables. In cases where a variable has been shown to predict both intention and behaviour, a calculation can be undertaken to establish the overall effect that variable has on behaviour, taking into account its influence on behavioural intention and the direct effect on behaviour. The resulting output for each set of regressions is therefore a set of diagrams presenting the influence of independent variables on intentions and behaviour using an easily interpretable set of arrows, sized to represent their overall effects.

Within this research project, path analyses were undertaken in two stages. First, three path analyses were used to examine the influence of the variables in Figure 4.3 in explaining the three types of environmental practices described in

Chapter 6. Second, 12 path analyses were undertaken to examine the influence of these variables according to both lifestyle group and behavioural practice. This second set of analyses was undertaken as part of the DEFRA-funded research and forms the basis for both the quantitative data analyses in the second half of this chapter and the qualitative data analysis examined in Chapter 9 (Barr et al. 2006; 2007a).[1] The next section therefore presents the results from examining the three behavioural practices in the first instance.

Path Analyses 1: Explaining Environmental Practices

The whole data set of 1265 questionnaires was used to explore the factors that influenced purchase decisions, habits and recycling behaviours and are presented in Table 8.1. The table lists the variables and factors explored in Chapter 7 in the rows in the left-hand column (see Tables 7.1–7.4). The following three columns relate to the prediction of behavioural, intentions whilst the final three columns pertain to reported behaviour. The final three rows of the table provide statistics illustrating the variance explained by each model (R^2_{adj}), the significance of the model (F) in providing a good fit to the data and an associated p-value with this F statistic. As can be seen in Table 8.1, the degree of explanation is variable and this should be borne in mind when interpreting the following analysis. However, in psychological terms, the models are acceptable for an exploration of the factors involved in determining intentions and behaviours.

The path analyses, which enable the reader to understand the role of variables on both intension and behaviours, are provided in Figures 8.1 to 8.3. The path diagrams for each regression show the influence of the significant variables on behaviour in terms of beta weights, but this time indicated by the thickness of each arrow. Those variables that influence both intention and behaviour have individual beta weights alongside each path, with the overall effect on behaviour provided in the label box. This is calculated by multiplying the indirect effects (the Beta weights) and then adding the direct effect. Path analysis therefore provides a useful analytical tool to demonstrate how the conceptual framework can be used analytically to demonstrate the influence of particular variables.

Figure 8.1 provides the path analysis for purchase decisions behaviour. The level of explanation of purchase behaviour was the highest of the three behaviours identified (50 per cent). The first key point to note in Figure 8.1 is the large influence of intention on behaviour. This is in accordance with Fishbein and Ajzen's (1975) notion of behaviour change and illustrates the critical role of 'willingness' on reported action. The second point regarding Figure 8.1 is the complexity of factors that influence both intensions and behaviours. In terms of intentions, the dominant influences are psychological in nature, with

---

    1   The material presented in this chapter is based on work contained in an unpublished technical report to DEFRA (Barr et al. 2006) co-authored with Andrew Gilg and Gareth Shaw from the University of Exeter. The author wishes to acknowledge their work as part of this chapter.

Table 8.1 Multiple regression models for behaviour and behavioural intention according to the three types of environmental practices

Factor scale or regressor	Behavioural intention			Behaviour		
	Purchase decisions (intentions)	Habitual (intentions)	Recycling and waste (intentions)	Purchase decisions (behaviour)	Habitual (behaviour)	Recycling (behaviour)
Behavioural intention						
Purchase decisions				0.32		
Habitual					0.28	
Recycling						0.12
Situational variables						
Age		-0.13		0.14		0.13
Gender	0.08					
Income				-0.14	-0.18	
Education level				0.07		
Car ownership			0.07			
Kerbside recycling collection						0.09
Garden	0.33		0.08			
Number in household				0.08		
Membership of community groups				0.11		
Water butt						
Water saving device in toilet				0.11	0.11	
Water meter		0.15			0.09	

Table 8.1 cont'd

Factor scale or regressor	Regression model					
	Behavioural intention			Behaviour		
	Purchase decisions (intentions)	Habitual (intentions)	Recycling and waste (intentions)	Purchase decisions (behaviour)	Habitual (behaviour)	Recycling (behaviour)
Social and environmental values						
Limits to growth	0.1					
Psychological variables						
Concern and commitment			0.3			
Willing to sacrifice comfort		0.18			0.13	
Awareness of norm to help the environment		0.06		0.1		0.05
Environmental beliefs	0.12	0.22				
Trust and responsibility	0.1	0.22	0.09	-0.09		-0.06
Moral motives	0.2	0.22	0.18	0.1	0.18	0.08
Green consumer beliefs	0.17			0.26		
Influence of price	0.1	0.09				
Brand loyalty	-0.08					
Personal instinct	0.04					
Satisfaction in helping the environment		0.07				
Logistics and convenience			0.13	0.12		0.26
Environmental rights					0.08	0.07
R^2_{adj}	55%	37%	31%	50%	41%	49%
F	104.31	76.41	59.92	43.4	80.6	109.12
p	<0.001	<0.001	<0.001	<0.001	<0.001	<0.001

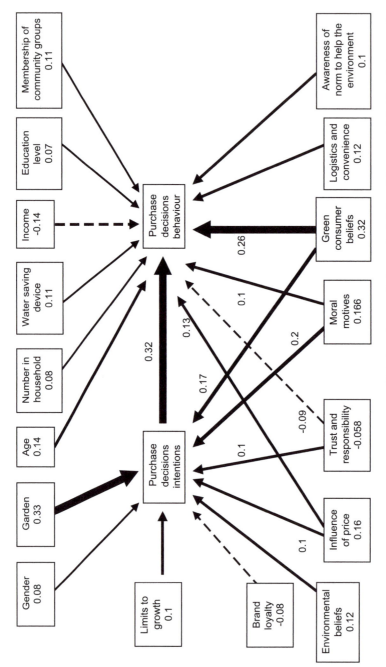

*Note: Arrows increase with size according to their predictive power. A dashed line indicates a negative relationship. Coefficients in boxes indicate the overall effect of that variable on purchase behaviour. Coefficients alongside arrows indicate the direct effect of the variable if there is more than one path from that variable.

Figure 8.1   Path analysis of purchase related behaviour

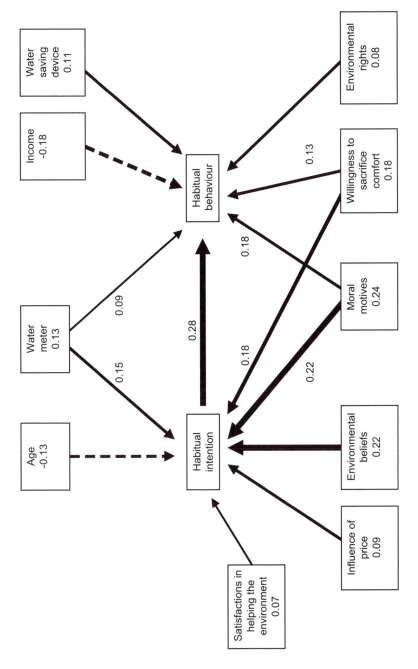

Figure 8.2 Path analysis of habitual behaviour

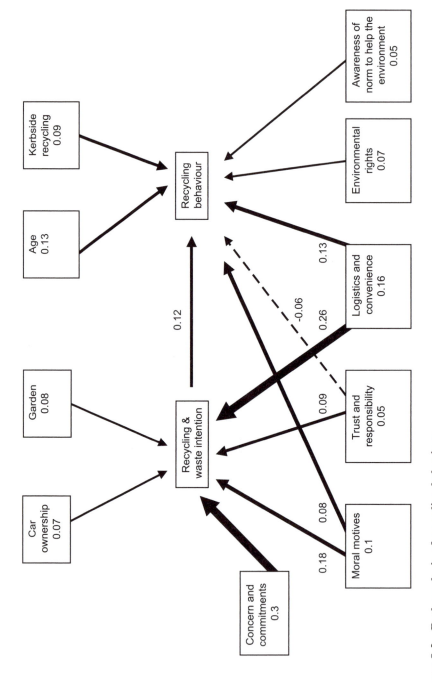

Figure 8.3 Path analysis of recycling behaviour

a combination of powerful effects from the influence of price to moral motives. Yet this variety of influences is in stark contrast to the singular role of gardens in predicting behavioural intentions. This result is intriguing and no immediate explanation is offered here, suffice it to state that an association between access to private green space and purchase behaviour may be reflective of home composting activities, which comprise part of the Purchase Decisions factor. Indirectly, it could be hypothesised that those with garden access may be more exposed to other pro-environmental activities, which are reflected in their purchasing behaviour.

Finally, behaviour has a greater number of direct influences from situational characteristics, although green consumer beliefs are the most dominant factors overall. Such a position implies that the complexity of defining purchase decisions is also reflected in the wide range of influences which seem to govern intentions and behaviours.

Figure 8.2 provides data relating to habitual behaviour. Compared to Figure 8.1, the number of variables influencing intentions and behaviour are significantly reduced. Indeed, the link between intensions and behaviour, whilst strong, is weaker than for purchase decisions. Habitual behaviours appear to be heavily influenced by the key psychological factors of environmental beliefs, moral motives and a willingness to sacrifice comfort, all indicating that undertaking habitual actions requires and represents key motives to act. However, it is also important to acknowledge the role of income, which demonstrates that those on lower incomes tend to be greater conservers of energy and water in the home.

Finally, Figure 8.3 illustrates that recycling behaviour is governed strongly by a perceived notion of the logistical ease and convenience of recycling. Although the presence of kerbside recycling does have a significant influence on behaviour, the combined role of perceived convenience is very important here. However, it is notable that the link between intentions and behaviours considerably reduced when compared to purchase decisions and habitual behaviour. Such a finding would imply that recycling is an activity which is not primarily determined by different intentions to act, but rather the logical ease and services provided. This is a positive finding implying that most individuals are predisposed to recycle, but require the facilities to do so.

These detailed path analyses present key findings that once again justify the decision to examine behaviour in terms of key environmental practices. The results reveal a range of factors that influence each set of practices noted for the differences in the variables that predict the three forms of behaviour. Purchase decisions, as the most complex and diverse of the behaviourally-defined practices, has a commensurate range of influences, spanning the three sets of influencing factors identified in Figure 4.3. Despite this range of factors, there are clear trends that emerge, highlighting the role that gardens play in influencing behavioural intentions and the key impact of green consumer beliefs. The link between intentions and behaviour was also the strongest for purchase decisions. Overall, these key findings indicate that the generally low uptake of purchase-related activities is reflected in the stronger link between attitudes to undertake such behaviours and actual action. In other words, purchase decisions require a commitment and attitudinal shift to effect change that is not as widely reflected for other practices. The influence of gardens is somewhat perplexing and may reflect the greater tendency of those who make green consumption decisions to have access to private green space, especially for composting, which situationally provides opportunities for such behaviour. Nonetheless, the importance of green consumer beliefs appears to impact on both intentions and behaviour and presents a clear link between positive attitudes (such as a willingness to pay more for environmentally-friendly products, concerns with food safety and health issues and the significance of buying 'locally') and behaviour.

In contrast, habitual behaviour is influenced by a far smaller set of variables, with key attitudinal constructs presenting the main motivating factors on both intention and behaviour. The belief that saving energy and water will have positive environmental impacts alongside the moral imperatives of saving energy and water, appear to have a very strong influence. Indeed, situationally, being on a lower income also appears to act directly to influence behaviour. Taken collectively, habitual behaviour is governed by a range of morally-driven factors.

In comparison to these higher-level beliefs, recycling behaviour appears to be mainly influenced by factors such as logistics, convenience and service provision. What is notable is the weak link between intentions and behaviour, implying that positive attitudes towards recycling do not necessarily transform into recycling behaviour, demonstrating the influence of key factors in driving this very specific form of behaviour.

These results provide useful data relating to the overall influence of a wide range of variables from the sample as a whole. However, in terms of their utility from a policy perspective, they present challenges in terms of delivering significant messages for moving policy forward. As such, these results are still presented at the 'population' level, with no means of creating and implementing practical instruments that can be applied to a specific audience or target segment. Accordingly, a series of path analyses were undertaken in which the three environmental practices were analysed according to lifestyle group. This resulted in twelve path diagrams. The next section examines these findings in detail to examine the role that lifestyle segmentation can provide in differentiating the results in Figures 8.1 to 8.3 according to different groups.

Path Analysis 2: Factors Influencing Lifestyle Groups

As with the broader analysis provided above, it is important to contextualise the following analysis with an exploration of the quality of each regression model used to construct each path diagram. Table 8.2 presents an overview of the multiple regression models for each environmental practice and according to each behavioural cluster. These are divided according to behavioural intention and behaviour. The explanations offered by the models are variable; recycling behaviour in Cluster 4 has a good explanation of 72 per cent for behaviour, whilst purchase behaviour for Cluster 3 only has an explanation of 24 per cent. Accordingly, reference to Table 8.2 will be helpful in evaluating the following path analyses. Figures 8.4 to 8.15 present the detailed findings from the path

Table 8.2   Multiple regression models for behaviour and behavioural intention according to the three types of environmental practices and four behavioural clusters

Statistics	Purchase behaviour							
	Cluster 1		Cluster 2		Cluster 3		Cluster 4	
	Intention	Behaviour	Intention	Behaviour	Intention	Behaviour	Intention	Behaviour
R^2_{adj}	55.2	38.9	57.5	31.5	59.2	23.9	80.3	65.6
F	57.0	41.57	42.4	19.4	75.6	17.3	26.8	12.13
p	<0.001	<0.001	<0.001	<0.001	<0.001	<0.001	<0.001	<0.001

Statistics	Habitual Behaviour							
	Cluster 1		Cluster 2		Cluster 3		Cluster 4	
	Intention	Behaviour	Intention	Behaviour	Intention	Behaviour	Intention	Behaviour
R^2_{adj}	38.8	31.1	35.6	28.5	38.5	23.3	54.1	71.8
F	28.3	25.5	30.3	18.8	41.2	18.3	10.2	11.7
p	<0.001	<0.001	<0.001	<0.001	<0.001	<0.001	<0.001	<0.001

Statistics	Recycling Behaviour							
	Cluster 1		Cluster 2		Cluster 3		Cluster 4	
	Intention	Behaviour	Intention	Behaviour	Intention	Behaviour	Intention	Behaviour
R^2_{adj}	38.6	27.9	25.0	26.5	31.3	44.6	55.7	71.5
F	33.8	11.6	12.1	15.4	22.6	29.0	14.9	15.6
p	<0.001	<0.001	<0.001	<0.001	<0.001	<0.001	<0.001	<0.001

analyses. As for the previous section, the results will be examined according to environmental practice, beginning with purchase decisions. However, the reader should note that to ensure the central message of each diagram is conveyed, the labels for the psychological variables have been altered to reflect their meaning more appropriately. This enables interpretation of the diagram without recourse to earlier tables. Accordingly, label wording does vary somewhat from that provided in Table 8.1.

Purchase Decisions

Figures 8.4 to 8.7 provide the results of the four path analyses for purchase decisions for each behavioural cluster. A cursory examination demonstrates that in contrast to the simpler analyses already undertaken for purchase decisions (Figure 8.1), there are large differences between the clusters. In more detail, Figure 8.4 shows that individuals within Cluster 1 (Committed Environmentalists) were positively motivated by environmental concerns, willing to pay more to help the environment, were willing to sacrifice comfort and felt a moral obligation to help the environment. Negative influences were household size, income and conservative values (implying that those who were less conservative, lived in smaller properties and on lower incomes were more willing to make sustainable purchasing decisions). For this committed group of individuals, green purchasing behaviour was therefore strongly influenced by a series of morally-motivated attitudes and concerns for the environment and, most notably here, this group was willing to expend more financial resources to pay for environmentally-sound products. They were also more likely to hold pro-social values and less likely to be conservative in nature. Overall, this committed group appeared to be highly influenced by their underlying values and attitudes, with a very strong link between intentions and behaviours.

Figure 8.5 shows that the behaviour of those in Cluster 2 (Mainstream Environmentalists) was strongly influenced by access to a garden and a combination of environmental concerns, green consumer beliefs and moral obligations. Compared to individuals within Cluster 1, this group was influenced by a far wider range of variables with the link between intentions and behaviour being weaker than for the committed group, although still very strong. Although this group was willing to pay more for environmentally-friendly products, a series of green consumer beliefs was more important than for the committed group, alongside moral and environmental concerns. However, the overwhelming influence was the access individuals had to a garden. This may indicate that those with access to private green space hold positive intentions to purchase green products, although it is notable that this does not necessarily imply that their behaviour is influenced by such access. Overall, this group had a series of psychological and situational bases to their intentions to buy green produce, with behaviour being influenced mainly by a willingness to pay, the satisfaction derived from buying green produce and green consumer beliefs.

Figure 8.6 shows that individuals in Cluster 3 (Occasional Environmentalists) were similar to those in Cluster 2 in terms of being strongly influenced by access

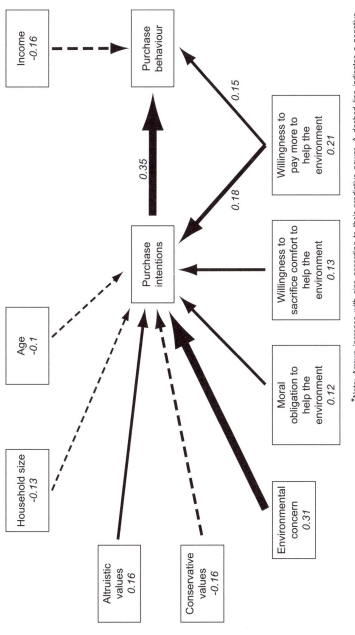

Figure 8.4 Path analysis of purchase related behaviour for cluster 1 (Committed Environmentalists)

Please note that the wording on some labels differs from previous diagrams to enable the reader to interpret the data without reference to other tables and diagrams.

*Note: Arrows increase with size according to their predictive power. A dashed line indicates a negative relationship. Coefficients in boxes indicate the overall effect of that variable on purchase behaviour. Coefficients alongside arrows indicate the direct effect of the variable if there is more than one path from that variable.

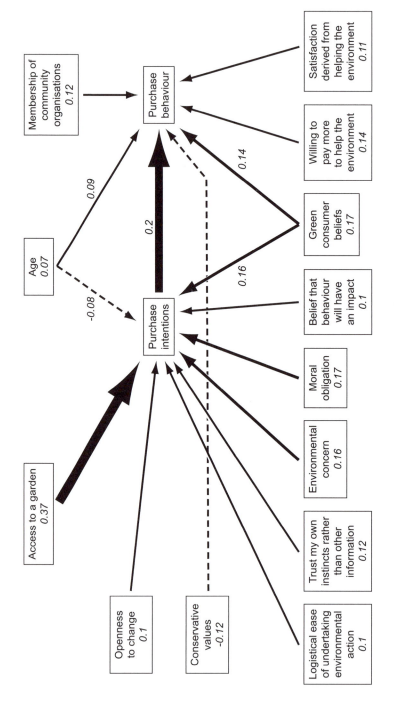

Figure 8.5 Path analysis of purchase related behaviour for cluster 2 (Mainstream Environmentalists)

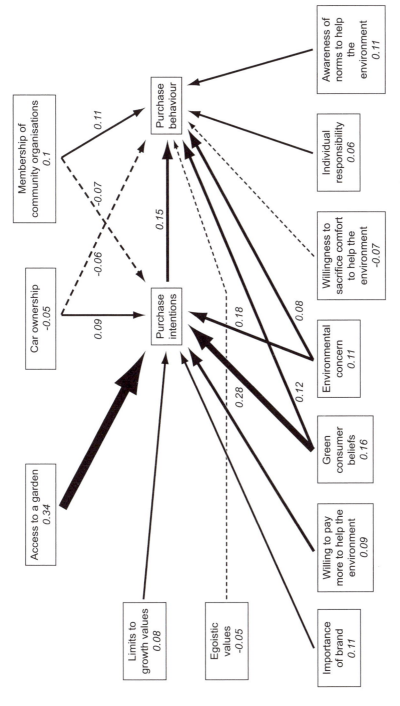

Figure 8.6 Path analysis of purchase related behaviour for cluster 3 (Occasional Environmentalists)

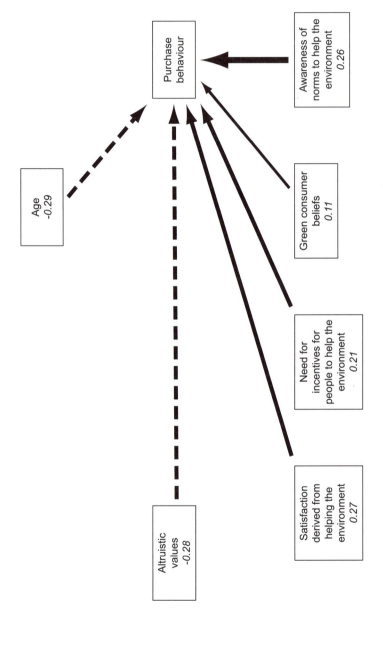

Figure 8.7 Path analysis of purchase related behaviour for cluster 4 (Non-environmentalists)

to a garden. However, the notable point from Figure 8.6 is the discord between intentions and behaviour. Whilst behavioural intentions were influenced by some of the factors that were seen for the previous two clusters (such as willingness to pay and green consumer beliefs), green consumer beliefs and environmental concern comprised more powerful factors in predicting reported purchase behaviour. Noting that most people in this group did not engage in green consumption behaviour, it appears that only those with a strong environmental concern were undertaking this type of activity.

Finally, Figure 8.7 shows that individuals within Cluster 4 (Non-environmentalists) were completely different from the other three clusters in that very few variables influenced their purchasing behaviour. Most notably, there were no significant influences on behavioural intentions at all, indicating that any intention-behaviour link was very weak indeed. However, the variables that did influence behaviour had a greater explanatory level. A strong positive influence was satisfaction in helping the environment, indicating that the very few in this group that did engage in green purchasing behaviour were strongly influenced by the sense of purpose and satisfaction they derived. Indeed, those with green consumer beliefs were also more likely to make green purchases. This group felt that they should be given incentives to help the environment, indicating that they were susceptible to extrinsic motives, as well as intrinsic motivations. The group were also heavily influenced by the behaviour of others with 'awareness of others' behaviour' being high. Negative influences were age (those being younger more likely to engage in green purchasing behaviour) and altruistic values (those with weak pro-social values more likely to help the environment). These results must be treated with some caution, given the small sample size and the few numbers of individuals who engaged in these practices. Nonetheless, this group did seem to be more concerned with the influence of others' behaviour and were the only group to highlight incentives as significant.

In summary, a complex and wide-ranging set of factors influenced decisions to participate in sustainable forms of purchasing behaviour. These focussed around psychological factors, although the situational influence of access to a garden was very important. Overall, social and environmental values exerted very little influence on purchase behaviours. There was a very strong link between intention and behaviour for the most committed groups, but for the Occasional Environmentalists (Cluster 3), there was a much weaker relationship between intention and action. Indeed, for the Non-environmentalist group (Cluster 4) there was no significant influence from behavioural intentions at all.

The most committed group (Cluster 1) demonstrated high levels of environmental concern and, with a strong intention-behaviour link, this influenced overall behavioural commitment. However, a willingness to pay more to help the environment was also important in mediating intentions into actions.

Although this factor was also significant for the Mainstream Environmentalists (Cluster 2), the intentions to purchase green produce for this cluster were overwhelmingly influenced by access to a garden, alongside the mediating factor of green consumer beliefs. Although access to a garden may be representative of individual circumstances, it does appear that for those likely to make decisions to help the environment within a retail context, access to a garden provided a motivator for positive behavioural intentions. Nonetheless, the mediating factor between intentions and behaviour related to a set of green consumer beliefs that emphasised the importance of health, safety and environmental concerns when buying products, alongside the importance of 'buying local'. Accordingly for this cluster, the situational influence of access to a garden, coupled with positive consumption beliefs were the most important factors influencing activity.

For the Occasional Environmentalists (Cluster 3), the picture remained very similar, with a strong influence from access to a garden on behavioural intentions. However, it is worth noting that the relationship between intentions and actions was less strong and therefore the influence of direct predictors of behaviour becomes more significant. Remembering that individuals in this cluster were less likely to undertake green purchasing behaviour, those that did were influenced by both underlying environmental concerns and green consumer beliefs.

Finally, Cluster 4 demonstrated that the few in the Non-environmentalist group that undertook green purchasing behaviour were guided by factors such as the influence of others' behaviour and the satisfactions they derived from green purchasing.

Habitual Behaviour

Figures 8.8 to 8.11 examine 'Habitual behaviour'. Once again, there are large differences between the clusters, although perhaps not as marked as for those of 'Purchase decisions'. Figure 8.8 shows that individuals in Cluster 1 had similar motivations as those for the 'Purchase decisions' type reported in Figure 8.4, notably in terms of moral obligations, a willingness to pay to help the environment and a willingness to sacrifice comfort. However, environmental concern has dropped markedly. There are no negative variables compared to those for 'Purchase decisions' and interestingly conservative values have moved from a negative to positive influence. Overall, this committed group appeared to be influenced by a willingness to sacrifice comfort to help the environment and a strong moral commitment that energy and water saving were morally desirably practices.

Figure 8.9 shows that individuals in Cluster 2 had a similarly wide range of variables influencing behaviour, as reported for purchase decisions in Figure 8.5. There was a strong link between the two behaviours (purchase decisions and habitual behaviour) in terms of 'Moral obligations' and 'Satisfaction derived' (see below). However, new factors become evident for habitual behaviour. In the first instance, intentions to undertake habitual behaviours were influenced significantly by the 'Logistical ease' of participating, which relates to the extent to which individuals felt the behaviour was simple and convenient to undertake. Similarly important is the 'Belief that actions will have an impact'. This implies that individuals in Cluster 2 who were willing to undertake energy and water saving behaviours were more likely to believe that their behaviours would have a significant impact. Behavioural intentions were also influenced by their 'Willingness to sacrifice comfort'. All of these psychological factors imply that

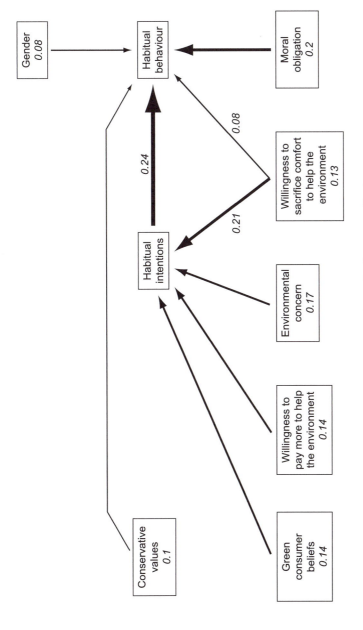

Figure 8.8 Path analysis for habitual behaviour cluster 1 (Committed Environmentalists)

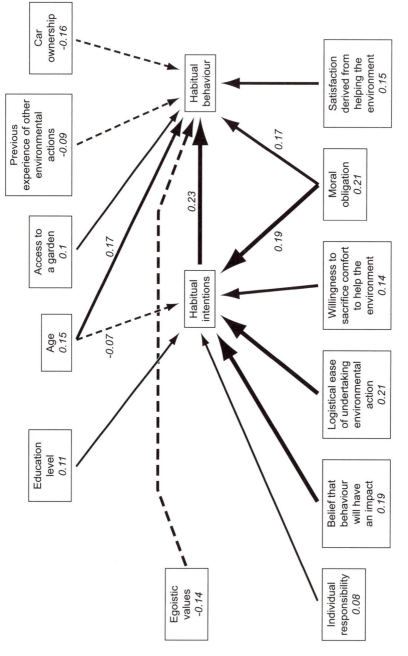

Figure 8.9   Path analysis for habitual behaviour cluster 2 (Mainstream Environmentalists)

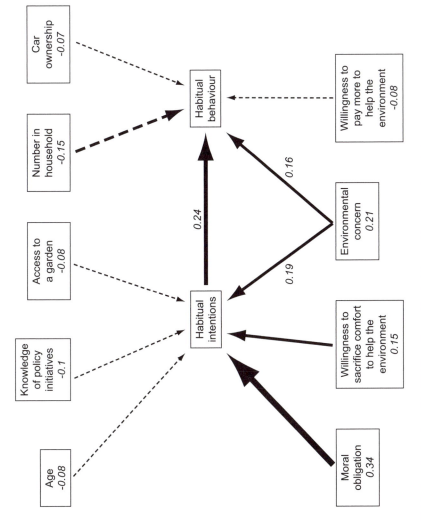

Figure 8.10 Path analysis for habitual behaviour for cluster 3 (Occasional Environmentalists)

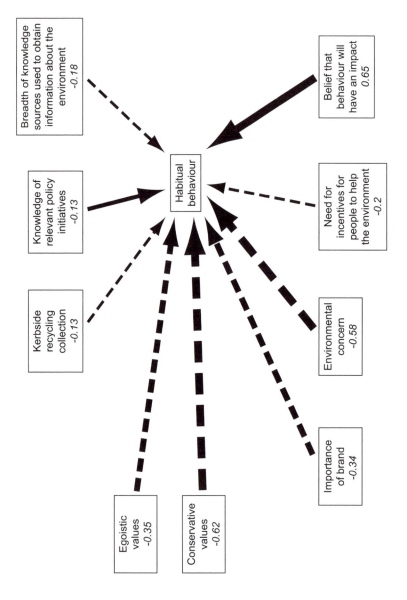

Figure 8.11 Path analysis for habitual behaviour for cluster 4 (Non-environmentalists)

the intention to save energy and water was influenced by a range of positive attitudes towards these activities. However, although the link between intentions and action was moderately strong, a large number of variables influenced either intentions and behaviour, or behaviour alone. The moral obligation to undertake a behaviour was once again significant, as had been the case for purchase decisions, demonstrating the effect that moral motives have in influencing the behaviour of this group. Indeed, this cluster also derived a number of satisfactions from their behaviour. Situationally, age had a positive influence overall, with those who were older tending to save water and energy more often. It is also interesting to note that those who did not own a car were more likely to engage in habitual behaviours. These findings indicate that this cluster is significantly influenced by underlying attitudes and moral motives to save energy and water in the home, with a moderately strong link between intentions and reported behaviour.

Figure 8.10 shows that individuals in Cluster 3 have far fewer influences on their 'Habitual' rather then their 'Purchase decisions' (Figure 8.6). In terms of behavioural intentions, 'Moral obligation' emerges as a new and strong influence for this cluster, being the most significant influence on intentions. 'Willingness to sacrifice comfort' also exerts a moderate influence on intentions to save energy and water. Although there are some situational influences, the influence of these is markedly smaller than for the psychological factors, with social and environmental values exerting no significant influence at all. As for Cluster 2, the intention-behaviour relationship is moderately strong. However, when compared to Figure 8.7, the influence on both intentions and behaviour of 'Environmental concern' becomes much stronger. Indeed, situationally, smaller households tended to save more energy and water. These findings indicate that this cluster has underlying moral obligations and a willingness to sacrifice some comfort, which are realised in terms of the transference from intentions to behaviour by the influence of environmental concerns, which mediate this critical relationship. Once again, this cluster demonstrates the utility of examining the intention-behaviour gap as a vehicle for exploring the alternative influences on reported behaviour rather than stated intentions.

Finally, Figure 8.11 shows a wider range of variables influenced habitual behaviour for Cluster 4 (Non-environemntalists) than for purchase behaviour. There is a striking difference between the variables that influence habitual behaviour in this cluster and the major difference between this cluster and those shown in Figures 8.8 to 8.10 is that there were no significant factors that influenced behavioural intention. Accordingly, the discord or 'distance' between intentions and actions was so great that all significant variables were found to influence behaviour only. As can be seen from Figure 8.11, the individuals in Cluster 4 who were likely to undertake energy and water saving behaviours were highly influenced by social values, with individuals demonstrating pro-social and open values (i.e. the opposite of egoistic and conservative values). Unlike Clusters 1 to 3, this demonstrates that these individuals were highly motivated by their underlying values and, given the few individuals in this cluster who did undertake such activities, it is interesting to note the strong influence of values alongside psychological factors. The influence of environmental concern was inverse, implying that those with lower levels of environmental concern were more likely to undertake habitual behaviours, although it is probable that this relates more to the lack of efficacy of this factor. More important are beliefs that the behaviour would be effective and the knowledge of policy initiatives. Overall, the few individuals in this cluster who undertook habitual behaviours were strongly influenced by a strong set of social values and beliefs that behaviour will be effective.

In summary, the results for habitual behaviour suggest that for Clusters 1 to 3, there were a reasonably specific set of factors that influenced habitual behaviour, focussed around psychological factors, which included moral obligations, a willingness to sacrifice comfort and environmental concerns. Depending on the cluster, these mostly influenced behavioural intentions with one mediating variable that transformed positive intentions into actions. For example, for Cluster 2 the mediating variable was moral obligation and for Cluster 3 it was environmental concern. These findings suggest that habitual behaviour is significantly influenced by psychological factors. In contrast, the Non-environmentalist group that comprised Cluster 4 had an alternative set of influences. They were strongly influenced by pro-social and open social values, with a strong belief that their behaviours would have an impact. They were not particularly concerned with environmental issues and did not have knowledge of policy initiatives. However, as noted previously, the fact that there were no statistically significant predictors on behavioural intention indicates that individuals in this cluster were only likely to undertake habitual activities if they were strongly influenced by their values and beliefs, rather than their underlying environmental concerns, willingness to sacrifice comforts or moral obligations.

Recycling Behaviour

In relation to the third behavioural factor, Figures 8.12 to 8.15 show the path analyses for 'Recycling Behaviour'. Recourse to Chapter 7 highlights the fact that recycling behaviour was frequently undertaken by all groups apart from Cluster 4 (Non-environmentalists). Indeed, it is also worth noting that recycling behaviour is a very distinctive activity, being heavily regulated and structured in many locations. Bearing these issues in mind, Figure 8.12 shows that for Cluster 1, there was a weak link between intentions and behaviour. Behavioural intention was heavily influenced by both environmental concerns and a belief that the behaviour would have an impact. However, behaviour was significantly influenced more by the access respondents had to a kerbside collections and the logistical ease of undertaking recycling, rather than behavioural intention. Accordingly, this committed group of environmentalists who recycled with great regularity were heavily influenced by the provision of recycling services and the logistical ease of recycling, as opposed to underlying values and attitudes.

Figure 8.13 shows that individuals in Cluster 2 were influenced by a range of variables. Bearing in mind once again that these individuals recycled with considerable frequency, the link between intentions and behaviour was still weak. A range of factors influenced behavioural intentions, involving both situational,

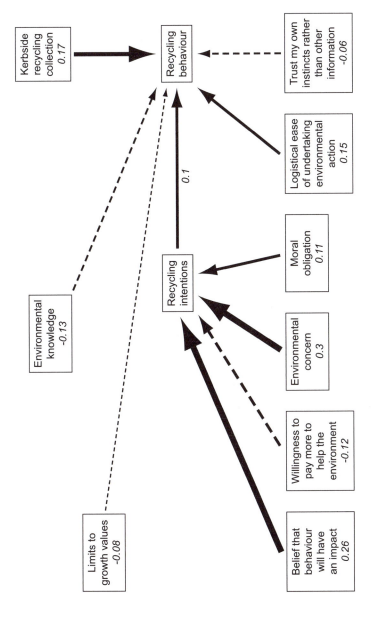

Figure 8.12 Path analysis for recycling behaviour cluster 1 (Committed Environmentalists)

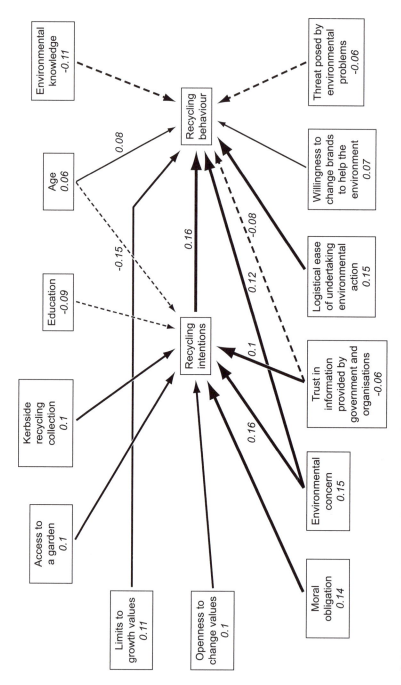

Figure 8.13 Path analysis for recycling behaviour cluster 2 (Mainstream Environmentalists)

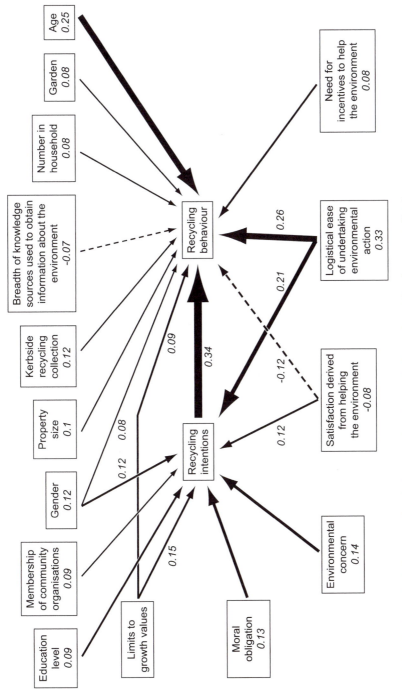

Figure 8.14 Path analysis for recycling behaviour for cluster 3 (Occasional Environmentalists)

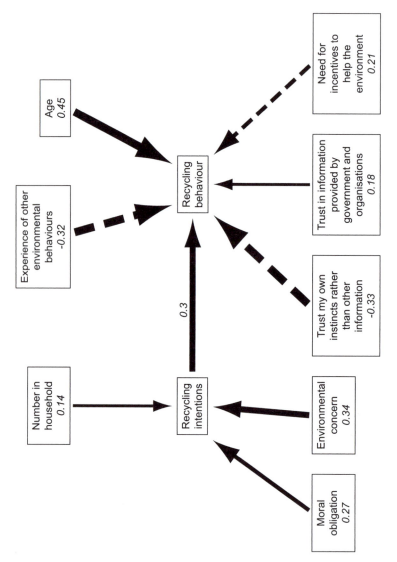

Figure 8.15 Path analysis for recycling behaviour for cluster 4 (Non-environmentalists)

psychological and value-based variables. However, these intentions were not strongly related to behaviour, which was once again influenced by the logistical ease with which respondents perceived they could undertake the activity.

Figure 8.14 presents a very different picture for individuals in Cluster 3. This group did recycle, but less frequently than for the Committed or Mainstream Environmentalists. The link between intentions and actions was very strong with a range of situational variables influencing intentions to recycle including education and gender. Indeed, environmental concern and moral obligations were key psychological factors. Accordingly, intentions to recycle appeared to have a larger number of influences than for other clusters. However, the mediating factor between intentions and actions was once again the perception that recycling was logistically simple. This, combined with a positive impact of age, appeared to most readily explain recycling behaviour for this cluster.

Finally, Figure 8.15 shows the influences on individuals within Cluster 4. Again this cluster has the fewest influences, but they are also much more significant as predictors. The link between intentions and behaviour is strong. For this group of Non-environmentalists, environmental concern and moral obligations were very significant in driving intentions to recycle. Behaviour was driven by intention, but also by a strong tendency for older individuals in this cluster to recycle more frequently, alongside trust of key organisations. What is interesting with regard to behaviour is the impact that the experience of undertaking other environmental behaviours had on recycling. Those who had less experience of other environmental actions were more likely to recycle. Bearing in mind the low levels of behaviour reported in this group, this indicates that those who recycled were unlikely to have engaged in any other forms of environmental behaviour. Overall, recycling for this group was dominated much less by logistical concerns and driven more by moral obligations and environmental concerns.

In summary, recycling behaviour is very distinctive. For Clusters 1 to 3, although key issues of moral obligation and environmental concern served to influence intentions to recycle, the mediating factor was the logistical ease with which the behaviour could be undertaken. For the Committed Environmentalists (Cluster 1), attitudes bore little relation to behavioural commitment, with a weak intention-behaviour relationship. This was also the case for the Mainstream Environmentalists (Cluster 2). For Cluster 3 (Occasional Environmentalists), there were a far greater number of situational influences, but these were mediated by the logistical ease of undertaking recycling. Only with regard to the Non-environmentalists (Cluster 4) was there a reliance on other factors. Accordingly, recycling represents a series of activities which are influenced by the logistical simplicity of participating. For those who do not participate, other factors are significant, focussing around environmental concern and moral obligations. These findings suggest that individuals in Clusters 1 to 3 were already morally and attitudinally committed to recycling and the main differentiating factor relates to the perception that recycling is simple to undertaken. In contrast, those individuals in Cluster 4 who did recycle were doing so on the basis of an environmental and moral commitment.

Conclusion: Segmenting the 'Value-Action' Gap

The conceptual framework provided in Figure 4.3 presents a means for understanding the influences that act on environmental behaviour. Even in Figure 4.3 these influences demonstrate the complexity of the environmental action, yet this chapter has exposed the multilayered nature of environmental action at the beginning of the twenty-first century in Britain. Chapter 5 explored the factors that researchers have exposed as being significant in the predictions of environmental action as seen from a compartmentalised perspective (i.e. energy saving, waster conservation, green consumption and recycling studies). This research has presented a series of (alternative) factors that influence different forms of environmental action. Through a systematic and logical exploration of firstly environmental practices and then environmental lifestyle groups, this chapter has explored the connected nature of this wide range of factors in influencing a combination of practices according to lifestyle groups. This has enabled a progressively nuanced understanding of environmental action and has highlighted some significant benefits for segmenting the population in order to understand attitudes and behaviours towards sustainable development practices.

Explaining Environmental Practice

The data provided in the first empirical section of this chapter (focused around Figures 8.1 to 8.3) highlighted the complexity of certain types of environmental action. In particular, purchase related activities demonstrated a range of factors spanning situational and psychological factors, alongside environmental values. Overall, the link between intention and behaviour was strong, with intention being influenced by access to a garden. The transference of intentions into behaviour highlighted the importance of both a willingness to pay more for environmentally-friendly goods, as well as a series of green consumer beliefs. This accords well with research by Roberts (1996) that has highlighted the role of attitudes towards green consumption and the influence of price in determining a commitment to undertaking green purchasing. Nonetheless, the importance of access to a garden was not a factor highlighted in previous literature, given the focus on psychological rather than situational factors. The importance of access to a garden therefore provides a key question for researchers in terms of the influence that access to private green space has on intentions to help the environment. One explanation may relate to the opportunities that green space affords for undertaking activities such as composting. Although these may not be reflected in actual behaviour, intentions may be enhanced by spending time within this particular spatial environment.

This picture becomes more complex when examining the differences between behavioural clusters, as evidenced in the second empirical part of this chapter. When purchasing behaviour is explored at this level it is clear that Committed Environmentalists' behaviour was driven mainly by environmental concerns and a willingness to pay more to help the environment, whereas both the Mainstream and Occasional groups were influenced far more by access to a garden and specific green consumer beliefs. The Non-environmentalist group, in contrast, was motivated by the normative expectations of their behaviour and any satisfaction they derived from the activities, representing a pro-self attitude.

This differentiation between the general analysis of environmental practise and the specific examination of behavioural segments also emerges with regard to habitual behaviour. Figure 8.2 demonstrated the significance of environmental beliefs, moral obligations and a willingness to sacrifice comfort to help the environment. In contrast, examination of Figures 8.8 to 8.11 demonstrated that committed individuals were more likely to engage in habitual behaviour if they were willing to sacrifice comfort to save energy and water and felt an underlying moral obligation to do so. Although this was partly evident within the Mainstream Environmentalist group, this cluster was also concerned with the satisfactions they derived from the behaviour. Occasional Environmentalists, who were less likely to undertake these activities overall, were motivated more by environmental concerns, whilst the few individuals in the Non-environmentalist group who undertook such behaviours were strongly influenced by values and the belief that their behaviour would be effective. In contrast to the literature on green consumption, these varied findings imply that the contested nature of the literature on energy and water saving, in particular between situational and psychological factors and the role of values (Stern 1992b), can be accounted for in some part by the alternative sampling strategies employed by researchers, given the varying results reported here.

Finally, recycling behaviour was clearly driven by logistical factors for both the general analysis of behaviour (Figure 8.3) and for Clusters 1 to 3 (Figures 8.12 to 8.14). Nonetheless, the fourth cluster, who were the least likely to recycle, were driven more by wider and deeper notions of moral obligations and environmental concerns. Once again, the varied findings related to recycling behaviour (Schultz et al. 1995) indicate that one potential reason for these discrepancies is the sampling strategy employed by researchers, who may have found differing influences (e.g. values or psychological factors) emerged with differing power according to the target population they were surveying.

Lifestyle Groups and Explanatory Analysis

This research has identified the influences that act on the four lifestyle groups that were identified in this study. Caution must be exercised when comparing the results to other research, given the segmented nature of the sample. Indeed, what this chapter has highlighted is the importance of lifestyle groups in determining the alternative influences on behaviour. It is possible that the contested nature of environmental behaviour research that has been identified as a problem since the 1980s (e.g. Hines et al. 1987) may partly be due to the methods used to design research projects, which have not sought to take into account the potential for population segments to display alternative attitudinal and situational characteristics.

The same qualifications apply to this research, given that a specific population (in Devon, UK) was sampled and then segmented. However, if we accept this qualification, some interesting conclusions emerge from the findings in this chapter. First, the title of this chapter focuses on the 'value-action' gap, or more theoretically, the intention-behaviour gap. The data presented in this chapter provide evidence that the 'distance' or discord between intentions and actions can often be variable and that in some cases intentions do not relate to actions at all. In many cases, individuals displayed very positive intentions, underlain by a series of positive attitudes, to undertake specific forms of behaviour. However, they were not always replicated by a strong link between intentions and behaviour, implying that even though individuals were positively disposed towards helping the environment, in many cases they did not, due to mediating factors, such as a willingness to pay for environmentally-friendly products or beliefs about green products (such as their health and safety benefits). This reinforces a central message of this text, which has emphasised the need to focus on specific factors that influence behaviour, rather than an assumption that increasing awareness and providing information will change behaviour through a linear process of attitudes and behaviour change.

A second point to emerge from this chapter relates to the alternative lifestyle groups that were identified. For all three sets of environmental practices, similar sets of variables appeared to influence the second and third clusters (Mainstream and Occasional Environmentalists), with the Committed and Non-environmentalist groups demonstrating distinctive properties. For example, the Committed group were those most willing to pay more sacrifice and comfort to help the environment. The Mainstream and Occasional groups relied more on green consumer beliefs and moral obligations to transfer positive intentions into behaviours, whilst the Non-environmentalists relied on values and their own experiences to mediate behaviour. Overall, these findings suggest that the Committed group were the moist willing to change their behaviour away from the status quo, with the Non-environmentalist group being the least willing to sacrifice their own lifestyles unless they derived specific satisfaction from helping the environment.

These differences between the lifestyle groups relate to a third key implication from the findings in this chapter. These data and those contained in Chapter 7 do provide some evidence for differences between lifestyle groups. As noted in Chapter 7, these are often graduated between groups, with Cluster 4 (the Non-environmentalists) being the only cluster to emerge as being very different. This is also borne-out with the data in this chapter. Nonetheless, the transference of these very broad findings into a practical context is more problematic. It will be evident how the data in Figure 8.1 to 8.15 can be utilised to contribute to social-psychological debates concerning an understanding of behaviour, but the data provided in both Chapter 7 and this chapter leave a number of questions unanswered. The main question relates to how findings on a conceptual level can be transferred for lifestyle groups into potential policies for behaviour change. For example, the data in this chapter reveal that Non-environmentalists who do help the environment on some occasions derive personal satisfaction from their behaviour and held more pro-self values, whereas Committed individuals were more willing to sacrifice comfort and pay more to help the environment, whilst holding more altruistic values. Whilst interesting, it is not immediately clear how such findings could be interpreted and transferred into a meaningful set of policy instruments targeted at these groups.

To overcome these potential difficulties, the research team decided to examine the findings from the quantitative element of the research within the context of a series of qualitative focus group discussions with individuals selected from each of the four segments, to explore both the quantitative finding and examine the potential for key policy changes for different behavioural segments. Central to this approach was a recognition that each segment's barriers and motivations for helping the environment were a critical part of encouraging behaviour. This focus on removing barriers and providing motivations for behaviour is very broadly termed 'social marketing' and it is to this topic that we now turn in Chapter 9.

PART 4
Applications

Chapter 9

Changing Behaviour:
A Social Marketing Approach

Introduction: Theory, Policy and Social Marketing

The findings presented in Chapters 6 to 8 provide an excellent basis on which
to provide a theoretically informed and original contribution to the debates
surrounding the conceptual bases of environmental action. These have
focused around three key conclusions that have emphasised the importance of
conceptualising environmental action in terms of environmental practices, lifestyle
groups and the identification of barriers and motivations to action. However, as
noted in Chapter 8, the transference of these theoretical and conceptually useful
findings into a policy context provides researchers with a significant challenge.
Concepts which can readily be identified as important influences on behaviour
such as the lack of convenient access to services or the belief that one's own
behaviour will not have an impact cannot be readily removed from the abstract
research context and placed within a specific policy framework.

This break in the link between research outputs (which can often be very
conclusive) and political reality raises what we can colloquially phrase the 'so
what?' questions amongst many policy-makers. If the results of research focused
around a topic such as environmental action cannot be applied with specific
policy instruments, is there any reason for policy-makers to engage with such
work? Part of the difficulty with the translation of research findings into policy
emanates from the academic goals of research, which are focused around testing
or applying a particular theoretical construct and have policy implications
almost as a convenient 'bi-product' or secondary objective. Many questionnaire
items are therefore constructed to compare directly to other studies and lack
the specificities that would be required for work that was solely focused around
changing policy.

So is there any potential to transform academic results into meaningful policy
implications? This chapter will seek to demonstrate how this may be achieved
within the context of behaviour change, providing the final empirical component
of the text (Figure 6.8). The effectiveness of a technique termed 'social marketing'
will be explored within a set of focus group discussions. Social marketing is a
method that attempts to identify specific barriers and motivations for undertaking
specific activities and then seeks to work with a target community to both
remove barriers and enhance motivations (McKenzie-Mohr 2000). This broad
definition of social marketing seeks to emphasise the importance of working at

the community level and to focus in on very specific elements that are preventing behaviour change. As McKenzie-Mohr (2000) notes, this has more often been applied within the context of health care, where specific segments of the population have been targeted with tailored policies to reduce regressive behaviours such as smoking or alcohol abuse. However, authors such as Everett and Pierce (1991–92) have outlined how community based social marketing techniques can be utilised to promote recycling behaviour.

The use of social marketing appeared to be a logical way forward in terms of ensuring the policy relevance of the research. The social marketing approach emphasises three key elements that were also central to the research reported in Chapters 6 to 8. First, there must be a focus on a specific behaviour for change to be effective. As noted in Chapter 6, the definition of specific behaviours related more closely to everyday forms of environmental practice is more likely to yield behaviour changes as they reflect everyday routines. Second, the notion of segmentation accords with the recommendation from social marketers that policies need to be focused on particular target groups in a community. Third, the path analyses in Chapter 8 illustrate the importance of highlighting the barriers and motivations for undertaking different behaviours. Accordingly, having applied the principles of social marketing within the quantitative research, the remaining issue was how to apply these findings within a policy context. To operationalise this, it was decided to undertake a series of focus groups to examine how the concepts explored by quantitative data could be used to examine policies with focus group participants. It was hoped that the application of focus group research could yield more contextually specific policies for changing behaviour.

A more specific and meaningful approach to social marketing is applied within the current research project. Although the essential element of McKenzie-Mohr's (2000) definition is central to the work reported here, 'social marketing' is framed within two critical contexts. In the first instance, the research was specifically interested in the role that social influences can play in promoting behaviour and therefore the potential that social marketing has for encouraging behaviour change through utilising normative processes within a community setting. This has already been illustrated in a number of behavioural contexts, such as Everett and Pierce's (1991–92) study of the impact 'block leaders' (an American term, which would translate as 'neighbourhood leaders' in the UK) could have on recycling behaviour. Indeed, within a British context the environmental charity Global Action Plan has utilised the notion of 'environmental champions' in businesses to encourage environmental behaviour (Global Action Plan 1999). Second, the research sought to explore the potential for 'marketing' to make a contribution to the behaviour change debate. A more specific definition of 'social marketing' would highlight the importance of marketing specific behaviours to target audiences, a common technique amongst those within the commercial sector, but one that has yet to be fully applied within the public sector. Increasing attention has been paid by many consumer industries to marketing around specific brands. A brand is any name, design, style, word or symbol that help distinguish a particular product. These can create value for customers in terms of both reducing search costs by identifying one product as being different and offering an implicit assurance of quality. To date, these ideas have been largely restricted to specific products and companies rather than abstract concepts like sustainable development. This research sought to assess the degree to which sustainability behaviour could be marketed as a brand and the extent to which this could be used to encourage behaviour.

Accordingly, this chapter will outline how the research developed a methodology for examining the potential for social marketing (both in its widest and narrower contexts) to deliver useful policy implications. The chapter begins with an outline of the methodological approach developed to apply these ideas and then provides a detailed and thematic examination of the qualitative data collected from the research. Finally, the chapter concludes by demonstrating the utility of this approach by providing a section dedicated to key policy implications.

Social Marketing Methodology

A qualitative research strategy was adopted in order to translate the potential offered by identifying lifestyle groups into potential recommendations for policy-makers (see the final column of Figure 6.8). This was the subject of a Department of the Environment, Food and Rural Affairs (DEFRA) funded research project during 2005–2006 (Barr et al. 2006; 2007a; 2007b).[1] This research was specifically funded to examine the role of social marketing in promoting behaviour change and was focused around developing the quantitative results from the ESRC-funded research into meaningful policies for changing behaviour. The focus groups were convened to reflect the target segments identified in Chapter 7 of the research. Accordingly, two focus groups were convened to represent each of the following lifestyle groups:

- 'Committed Environmentalists';
- 'Mainstream Environmentalists';
- 'Occasional Environmentalists';
- 'Non-environmentalists'.

During January of 2006, a market research company from the University of Exeter was commissioned to recruit and run the focus groups. Recruiting accurately to each of the clusters was critical to the success of the research yet the differences between the groups in terms of their behavioural attributes caused significant challenges for the researchers undertaking recruitment. For example, those individuals in Cluster 1 were distinguished from those in Cluster 2 by their propensity to compost. In both cases, individuals within these clusters undertook a wide range of purchase-related, habitual and recycling behaviours. Recruiting

---

1   The material presented in this chapter is based on work contained in an unpublished final report to DEFRA (Barr et al. 2007a) co-authored with Andrew Gilg and Gareth Shaw from the University of Exeter. The author wishes to acknowledge their work as part of this chapter.

for the 'Occasional' and 'Non-environmentalist' groups was simpler, given the greater differences in behavioural commitment between these clusters.

The recruitment was undertaken in Exeter High Street and at other locations in the city where individuals in specific clusters might be found, such as the local civic recycling centre. The recruitment questionnaire used a standard format and respondents were paid £25 to cover their expenses for attending the groups. The focus groups were all held in a suite at the Southgate Hotel in Exeter and lasted for one and a half hours, with refreshments being provided. All of the groups were tape recorded and fully transcribed.

The groups were taken through a discussion guide by the moderator who focused on an examination of the following three behaviours identified in the quantitative work:

- purchase decisions, e.g. shopping habits;
- habitual behaviour e.g. turning off lights, taking a shower rather than bath etc.;
- recycling behaviour, e.g. mainly waste management.

These three behavioural topics were used to frame the discussion around which the focus groups were given seven key issues to discuss:

1) general environmental attitudes and awareness;
2) recycling and conservation behaviour;
3) consumption attitudes;
4) attitudes towards branded products;
5) intended behaviour and actual behaviour;
6) policy-making and behaviour;
7) description of environmental attitudes.

These were directly related to both the findings as reported in Chapters 7 and 8 and specific policy-related objectives for this specific project. As can be seen, there was a preliminary discussion of environmental attitudes and awareness to generate discussion about environmental issues. The groups then moved on to examine the very specific activity of recycling behaviour, a topic which all groups could easily discuss given the proliferation of recycling services and policies. It was also felt that on the basis of the quantitative results, recycling merited a specific 'slot' given its distinctive nature. The groups then moved on to examine 'conservation' attitudes, which were related to the 'habitual' activities such as energy and water conservation in the home. Green consumer attitudes (relating to 'purchase decisions') were the final set of behaviours to be dealt with as a means by which to move the discussion onto the topic of the influence of marketing and branded products in the formation of environmental behaviours. The discussion then turned to a central element of the research identified in Chapter 8, notably the 'intention-behaviour' gap. This part of the groups' discussions examined the extent to which individuals could identify and account for a discord between their intentions and actions. Penultimately, the groups were then asked for their views concerning current policies for encouraging environmental behaviour, before finally being asked to identify themselves as one of the four lifestyle groups that had been derived during the quantitative research (interestingly, almost all individuals placed themselves into the groups identified previously).

The analyses of the focus group discussions were undertaken by a process of coding and then inter-group analysis through the use of a coding matrix. The research team found that six main themes emerged, broadly related to the seven key issues outlined above. These were:

1) awareness and responsibility towards the environment (Issues 1 and 7);
2) experience of actions and role of facilities (Issue 2);
3) cost and convenience (Issues 2 and 3);
4) branding (Issue 4);
5) intended and actual behaviour and barriers preventing behaviour (Issue 5);
6) awareness and influence of policy-making by controls and incentives (Issue 6).

Focus Group Analysis

The results from the focus group discussions will be described in turn, through an analysis of the four cluster groups for each of the six key themes.

Awareness and Responsibility towards the Environment (Issues 1 and 7)

The first part of the focus groups included a general discussion on the nature of environmental action and attitudes towards helping the environment. These exchanges covered a wide range of issues, although key themes that developed were the nature of collective action for sustainability and the notion of individual responsibility for the environment. All groups had an awareness of the key issues, such as recycling, but there were two key differences between the segmented categories. The 'Committed environmentalist' groups (focus groups 1 and 2) had the widest knowledge and one of these groups expressed concerns beyond local and everyday activities, using abstract concepts such as food miles and the 'duty to future generations' rather than practical examples to describe their attitudes and their sense of environmental responsibility. The need to foster more responsibility was clearly articulated by one participant who argued that:

> We need to get out of the mentality that I'm not going to do anything because nobody else does.

This problem of (a lack of) collective will was mentioned by both 'Mainstream' groups as well (focus groups 3 and 4). One group highlighted the 'snowball effect' of car use mitigating against public transport and 'forcing' people into their cars, while the other group argued that lifestyle changes would take a whole generation to evolve. Similarly, one of the 'Occasional' groups (5 and 6) showed quite wide awareness of the need for collective action, although the other 'Occasional' group expressed confusion over the range of facilities offered to help the environment and wanted more information, arguing that big business and the state should take the lead in changing behaviours.

The 'Non-environmentalists' (groups 7 and 8) were the only groups to express limited understanding of the issues. They focused on issues such as plastic bags in supermarkets and argued that the state should do more to provide them with information. One group argued that being environmentally friendly was 'uncool' and 'Swampy' was mentioned as a role model to be avoided, a reference to one renowned anti-road campaigner in the UK during the 1990s, with one participant stating:

> I[t] feel[s] really dodgy saying you're an environmentalist, sort of like you're Swampy and his mates. (Focus group 8, Non-environmentalist)

In conclusion, awareness was variable across the different environmental clusters ranging from a high level through to rather limited misunderstandings from the Non-environmentalists. Notions of responsibility also shifted according to cluster membership, with high levels of personal responsibility expressed by the more committed groups, with those less committed arguing that the state and businesses should become more involved.

Experience of Actions and the Role of Facilities (Issue 2)

This section of the group discussions focused on the more practical experiences of participants in undertaking specific activities. Most groups were well aware of what actions they could undertake but their take up was clearly influenced by psychological factors and situational variables. Most notably, composting was not popular because of 'poor' perceptions ('smelly' and 'dirty'), 'laziness' (unwilling to separate organic waste) and lack of facilities (notably internal space or the lack of a garden big enough to hide unsightly waste). These are common attitudes that are frequently expressed about recycling with which the reader is probably familiar.

The 'Committed' groups (1 and 2) emphasised the role of situational variables, eulogising the beneficial impact of water meters in making an explicit link between water saving and saving cash. The groups were concerned that labels on bins were not clear enough; for example, what types of plastic could be put into one bin or not. Finally, even these (committed) groups noted that environmentally-friendly behaviour still needed a 'Bit of an effort'. Nonetheless, these groups were clearly enthusiastic about helping the environment in most cases.

The 'Mainstream' groups (3 and 4) admitted that 'laziness' was often a factor in not always acting in an environmentally-friendly manner. Thus, devices which reduced toilet flushes were welcomed, but ironically some respondents thought that they abused showers by leaving them on too long in a perverse reaction to saving water from the bath they had forsaken. They also expressed concern regarding confusion about what could be recycled in kerbside receptacles. The

'Occasional' groups (5 and 6) claimed that they did most of the 'little things' well and that these had become second nature. Some participants noted that they felt a 'warm glow' when they acted well, but others acted erratically. Finally, the 'Non-environmentalists' (groups 7 and 8) supplied several reasons for failing to act in an environmentally-friendly manner. Most bluntly, one participant claimed that such actions were 'A pain in the arse' while others claimed laziness, expense and lack of space as reasons for failure to act.

In summary, most individuals acted in an environmentally-friendly manner some of the time and felt that the provision of facilities helped them to act in this way. However, inaction was easily engendered by slight difficulties, inconvenience or 'laziness'. Environmental habits were therefore 'easily learned' but just as easily 'lost in the heat' of day-to-day living. In this context it was clear that although the different clusters displayed alternative attitudes and behaviours, their actions were also conditioned by similar constraints. This can be illustrated by the following examples from participants in the focus groups to the difficulties of adopting an environmentally-friendly lifestyle:

- Committed environmentalists:
  - 'Inconvenience again really';
  - 'Well where you live, it can be inconvenient';
  - 'It needs to be made easier';
  - 'It's not practical yet is it?';
  - 'And it costs money';
- Mainstream Environmentalists:
  - 'Time and effort and awareness as well';
  - 'You want it to be environmental but also cheap';
  - 'I think another barrier is people not giving a damn and you think to yourself why do I bother';
- Occasional Environmentalists:
  - 'Yeah, I think it is very difficult, I think part of it is that you need to be more or less re-educated';
  - 'I think it boils down to awareness';
  - 'It's an inconvenience to start isn't it ... It's just that you have to put yourself out a lot to start with';
  - 'You kind of think is it worth me doing this?';
- Non-environmentalists:
  - 'Inconvenience';
  - 'I think its just habit as well. Normally when you throw something away it's just habit';
  - 'I think it's down to the fact that it's a lot more effort to act in an environmentally-friendly way'.

Cost and Convenience (Issues 2 and 3)

In a discussion ranging some of the most pertinent results from the quantitative study regarding the willingness to pay for environmentally-friendly products and the convenience of undertaking environmental action, all of the groups argued strongly that cost and convenience were key factors in acting in an environmentally-friendly manner, notably with regard to shopping in supermarkets and choosing the car rather than public transport. However, some respondents went out of their way to purchase local and/or organic products, sometimes claiming they were cheaper. In contrast, no respondents used public transport as an explicitly environmental act although there were some who supported congestion charging schemes.

The 'Committed' groups (1 and 2) did purchase local, organic or 'Fairtrade' goods as a matter of principle, but also because it was sometimes cheaper, although most respondents were willing to pay higher prices on principle. However, most shopping was still done by car and at the supermarket because of convenience and the availability of staple goods. The 'Mainstream' groups (3 and 4) expressed ambivalent views. For example one participant did not wish to 'pay through the nose' for organic products while another group member claimed that farm shops were cheaper, albeit less convenient. Some saw congestion charges as another tax and most noted that from the perspective of car use versus public transport, most people preferred to sit in their own cars because these were more convenient than public transport. The 'Occasional' groups (5 and 6) were also convinced that cost and convenience were major influencing factors, but one group argued that supermarkets and manufacturers could help by giving each product an environmental credit rating (such as a labelling scheme, currently used for food nutrition). This would enable consumers to 'trade-off' price with the environmental footprint of one product with another. Finally, the 'Non-environmentalist' groups (7 and 8) appeared to be the most influenced by price and argued that petrol prices were too high, although some respondents expressed a desire to buy locally if it was affordable.

In summary, price was a key determining factor for all participants but some individuals were willing to pay a greater amount for environmental goods out of principle. It should also be noted that individuals who said they would be willing to pay more often remarked that overall food and other goods was still too expensive. In contrast, convenience resonated strongly across all the groups and there was a tangible sense that apart from a small proportion of the participants, the dominant mode of shopping in supermarkets would continue and that the notion of using local food shops involving a greater time commitment would not be popular.

Branding (Issue 4)

The significance of branded goods was a major component of the group discussions used by the research team to assess the potential for using the notions of 'social marketing' as a tool for encouraging behaviour change. At the beginning of this chapter the notion of social marketing was introduced as a broad notion of how to positively encourage environmental action on the basis of dealing with specific barriers to participation. However, social marketing also involves the potential to 'brand' or market lifestyle choices at different market 'segments'. This was of particular interest to the research team who were interested in exploring how individuals related to notions of branding and the potential use of this for behaviour change policy.

As a concept, branding was perceived very differently depending on the type of product. Clothes and other non-food goods were seen as fashion items. Accordingly, some respondents bought branded goods to be in fashion, while others avoided such goods because they disliked the idea of being fashionable and sporting advertising labels. As one respondent from the 'Mainstream Environmentalist' group stated (focus group 3):

> I do buy some branded clothes but I wouldn't wanna buy something that had the brand printed all over it.

Views about branded foods were very polarised. Some believed that branded goods were the same as 'own label' goods, while others argued strongly that 'own label' goods were cheap and nasty compared to famous brands like Heinz. Quality was often mentioned as a factor in discriminating between branded and non-branded goods, most often for food but sometimes for non-food as well. In this context focus group participants stated:

> I do go for brands sometimes because they suggest quality. (Focus group 3, Mainstream environmentalist)

> It's just that brands that I buy, I know I'm going to get quality. (Focus group 7, Non-environmentalist)

A number of respondents claimed that labelling schemes were unreliable, using evidence supplied by friends who worked in the food industry. For example, the same batch of food being packed under different labels or free range eggs being substituted by battery eggs. A degree of cynicism was also revealed by this issue and raises the difficulty that, in a society like Britain where trust in politicians and advertising is waning, policy campaigns can encounter scepticism and apathy. However, some respondents reported that they would not buy brands perceived to use cheap labour in the developing world. As the following statements illustrate there were some variations according to lifestyle group. Thus, one respondent claimed:

> The brand would actually sometimes make me avoid certain products for example, with Nike and Gap, where the media has associated their brands with sweatshops. (Focus group 5, Occasional environmentalist)

In contrast another individual stated:

> In terms of brands I wouldn't be thinking about where it's made and sweatshops and all that. (Focus group 3, Mainstream environmentalist).

In summary, views about certain aspects of branding were only characterised by minor differences between the four groups with one fairly constant theme being the perception of quality and value-for-money that some brands had in favour of others or non-branded goods. There was limited evidence that niche environmental or ethical brands like 'Fair Trade' could break into the mainstream, although some respondents sought out such goods where they could. In terms of Fairtrade, most groups could identify the scope of these products but some of the Mainstream and Committed environmentalists raised questions about the environmental credentials of their production. Individuals across all the groups had made limited purchases usually for ethical reasons but many stated they were too costly. Nonetheless, what the Fairtrade brand does show is that as an ethical marketing concept it appears to have had a fairly wide influence across all the environmental clusters, lending some weight to the argument that branding may have potential within the environmental field.

Intended and Actual Behaviour and Barriers Preventing Behaviour (Issue 5)

The focus groups each spent a period of time discussing the perceived differences between what they intended to do and the realities of their behaviour. This was an important element highlighted by the quantitative research and the focus groups offered the opportunity to learn more about the differences between 'values and actions' within a colloquial context.

   All of the groups noted a big intention-behaviour 'gap', insofar as they often thought about how they could act more environmentally and then found reasons for not doing so. The 'Committed' groups (1 and 2) thought a great deal about the issues but were prevented from acting by a 'lack of choice'; for example, only the motor car could provide the transport they needed. Other issues highlighted by this cluster were cost and convenience or a common complaint about the overuse of plastic bags at supermarket checkouts. The 'Mainstream' groups (3 and 4) expressed similar concerns but also mentioned 'laziness' as a factor. Crucially one group expressed the bleak view that individuals acting alone can have little impact and thus by implication:

> Why should I bother to act environmentally? (Focus group 3)

The 'Occasional' groups (5 and 6) were divided into one group who 'Were not really bothered' (Group 5) and one that wanted to 'do better' and had good intentions, but who were prevented from doing so by impracticability, inconvenience and a lack of awareness of how they could act in an environmentally-friendly way. Finally, the 'Non-environmentalist' groups (7 and 8) highlighted reasons such as 'time' and 'effort' for not acting. A striking example was provided by one respondent who noted that on a long journey they started by driving at 70 mph in order to save fuel but abandoned the attempt after five minutes because other factors took over, for example time constraints and the lack of other people obeying the speed limit. This raises two crucial points:

- the creation of a culture which engenders low levels of response efficacy ('no-one else is driving at 60 mph, so why should I?');
- a perception of spending resources by time rather than by resource cost ('the priority is arriving earlier, not saving fuel or reducing emissions').

These points raise two policy challenges which are to:

- make the links between behaviour and consequences much more explicit;
- use such links to help change behaviour.

These issues are explored in more detail in the next section dealing with the implications of the research.

Awareness and Influence of Policy-making by Controls and Incentives (Issue 6)

The final part of the focus groups discussed the role of policy and the perceptions participants had of the role policy-makers played in changing behaviour. This part of the group discussions sough to ask participants about the views they held towards policy development for promoting behaviour change and, within the context of social marketing, the role of incentives (rather than purely penalties) for encouraging different lifestyle practices.

Awareness of policy measures tended to decline from Cluster 1 to Cluster 4 and most groups were against controls on individual actions, but in favour of incentives and controls on 'big businesses'. The 'Committed' groups (1 and 2) were aware of campaigns and policies but one group still wanted more publicity. Group 2 wanted controls but initially on business and industry, supported by education, whilst the other group favoured education rather than controls. Both groups highlighted the need for more incentives based on 'freebies' or better facilities. The 'Mainstream' groups (3 and 4) were both agreed that more publicity was needed but for different reasons. Group 4 wanted publicity to emphasise the level of threat caused by not acting in an environmentally-friendly way, whilst Group 3 wanted to improve awareness of how to act more responsibly. Both groups wanted better enforcement of existing controls, most notably littering.

The 'Occasional' groups (5 and 6) both mentioned a prize scheme initiated by the local council, which encouraged people to recycle goods and one group (5) emphasised the need to educate children into environmental awareness. Controls were favoured by both groups on 'big firms' and industry, with one group suggesting much higher environmental standards in new houses and the other arguing for taxes on big firms. Both groups highlighted the need for more incentives with one group favouring tax reductions for 'good' environmental actions.

Finally, both the 'Non-environmentalist' groups (7 and 8) were aware of the issues, but argued for different policies. One group thought the government should lead by example and argued that existing controls were not enforced enough, notably speed limits, but the other group was fairly cynical about campaigns to make them act environmentally, being against controls but in favour of taxes related to recycling rates. Both were not very convinced about incentives, arguing that these would be mostly ineffectual.

In conclusion, controls on individuals were not welcomed while incentives were favoured. Most groups also wanted large organisations to be more strictly controlled or to set a better example. These findings confirm the challenge of the behaviour-action gap identified in previous sections, illustrated by the notion that individuals 'want someone else' to be controlled but are pleased to received incentives personally. The environment is therefore very often seen as someone else's problem: either 'big' business needs to be targeted or 'someone else needs to give you an incentive'. The challenge remains, therefore, how most effectively to link day-to-day behaviours with abstract and long-term concepts like global warning.

Conclusions from the Focus Group Analysis

The focus groups expressed widespread awareness of environmental issues, but a reluctance to embark on substantial changes in lifestyle. This was particularly the case for the Non-environmentalist groups, who were unwilling to embark on major changes to their lives and aspirations. Accordingly, the main findings from the focus groups were that first, very few respondents were keen to adopt a radically different lifestyle, but were willing to make incremental adjustments. Such incremental change must also be seen within the context of incremental changes in related physical infrastructures for sustainability. Incremental change is just that, a steady movement towards a more sustainable set of lifestyle practices, but not a step-change. This was differentiated by lifestyle group, with those in more pro-environmental groups being more willing to make larger and longer-lasting changes to their lifestyles. Second, there is clear evidence of an 'intention-behaviour' gap, with individuals specifying many barriers to action, despite stating that they are willing to act. However, the 'distance' between intentions and actions is much wider for Non-environmentalists than for Committed groups, indicating that those more committed to the environment act more readily in accordance with their intentions. Third, specific barriers to participation need to be tackled, which vary across lifestyle groups, ranging from perceived inconvenience to lack of trust in both national and international authorities to act effectively. This weakens levels of response efficacy. Fourth, in response to this, personal responsibility needs to be tackled in relation to the ascribed roles attributed to the individual, the state and major companies. Crucially, there is a lack of response efficacy amongst individuals who ascribe a greater responsibility to external agents for 'being responsible' for the environment, before an individual behavioural commitment will be forthcoming. This can be framed at a range of scales and was expressed both in terms of a lack of national government commitment as well as at the international level. Fifth, related to this lack of response efficacy, discussion centred on the perceived role of 'big business' and in particular supermarkets and how modern lifestyles necessitate the use of such retailers that are perceived as being more unsustainable. Shifting away from this way of living was seen as problematic because of time and cost factors. Sixth, based on the previous two conclusions, behaviour change is most likely to occur at scales where levels of collective action can readily be engaged and measured, such as at the community level. The focus groups were all framed within discursive contexts where respondents used each others' behaviour as a measure of their own level of activity. Finally, the importance of economic factors were significantly highlighted by certain groups. There was a clear signal that although further environmental surcharges or disincentives would be unpopular, creating incentives to act in a more environmentally responsible manner would be effective.

The focus groups therefore provided excellent evidence that there were a number of key findings that could be applied within a policy context. The research team was then asked by DEFRA to examine the potential for creating a series of policy implications that could be directed specifically at current political initiatives for encouraging behaviour change. Accordingly, the final task of the research team was to utilise the focus groups and quantitative data to create a practical guide for changing behaviour in the form of a series of policy implications. The next section of this chapter provides a summary of these implications and acts as a way of demonstrating the utility for policy-makers of the approach adopted in both the ESRC and DEFRA-funded research projects and illustrates how a piece of academically and theoretically driven research can yield key policy implications for national and local government. As such, the next section represents the end point of the research process and (by definition) the beginning of a new agenda for research which will be outlined in the final chapter of this book.

Policy Implications of the Research

The implications from the research are divided into two key sections, relating to 'concepts' and 'implications'. The 'concepts' section explores the overall policy directions that the research highlights and suggests ways in which policy can be more effectively directed in strategic terms. The policy implications suggest more specific measures for policy, although it must be emphasised that these are still at a generic level and do not provide a 'practical guide' in the most literal sense. However, they are intended to highlight the key issues for policy-makers to focus on.

Concepts for Behaviour Change

Three concepts were identified which policy-makers need to take into account when designing behaviour change strategies and emanate directly from the quantitative and qualitative research. First, a focus on 'Practices, not problems': both previous academic research (see a review in Barr et al. 2005a) and the current study have confirmed that behaviours are 'practiced' within the daily routines of everyday life. Individuals who engaged in pro-environmental behaviour build pro-environmental activity into their purchase decisions (when shopping), their habitual daily routines (such as 'mundane' or seemingly inconsequential behaviour around the home, like the use of energy and water) and their recycling behaviour (the use and disposal of post-consumer wastes). The implication of this key finding was that new campaigns, at whatever spatial scale, need to address behaviour from a 'practice' perspective, rather than a 'problem-focused' viewpoint. The implication of this shift is therefore a move away from environmental campaigns focused on 'energy' or 'water' towards messages that promote reflection and action on 'purchase decisions' and 'being sustainable in the home'.

Second, a shift towards 'Targeting lifestyles': a further key concept to emerge from the research was that there needs to be a shift towards segmentation as a means by which to understand the alternative barriers and motivations to action for specific lifestyle groups. Segmentation is common place in social marketing practice (NCC, 2004) and seeks to address key barriers to change. The research has demonstrated that there are lifestyle groups that can be identified in relation to environmental practices and that the influences acting to predict behaviour are complex, dependent on the environmental practice in question. The key implication for policy, therefore, is that policy needs to be specifically nuanced in terms of both practices and lifestyles.

Third, an acknowledgment of the importance 'Social marketing' has to the behaviour change process: previous research (McKenzie-Mohr, 2000) has highlighted the role of social marketing as a technique for generating change. The key principles in this approach are focused around identifying key barriers to change and working at the community level to effect change. This research explored the ways in which social marketing as a broader concept of normative engagement in behaviour could be used to effect change. These three concepts will now be elaborated in relation to seven key implications for policy, driven by the findings from this research.

Implications of the Research Findings

The following are presented as key practical implications for policy emanating from this research and key quotations from the focus groups are used to illustrate points, although the implications are drawn from both the quantitative and qualitative stages of the research.

The Significance of Social Acceptability

> I[t] feel[s] really dodgy saying you're an environmentalist, sort of like you're Swampy and his mates. (Focus group 8, Non-environmentalist)

Implicit within the research findings was that very few individuals in the less committed groups were willing to make radical changes to their existing lifestyles. As noted in the first 'concept' previously, environmental action is adopted within the context of existing daily practices and adopting a radically alternative lifestyle can be perceived as being 'risky' in social acceptance terms. As the quotation above reveals from a respondent (who held generally pro-environmental *attitudes*), there was an unwillingness to adopt what they perceived to be a socially undesirable set of practices. The implication for policy-makers is that emphasis needs to be placed on effecting incremental behavioural shifts. As the third 'concept' discussed previously intimates, this is most likely to be effective when behavioural change is framed within socially 'desirable' contexts (see below). This emphasis is most effective when speaking to 'Non-environmentalist' groups. In contrast, those in more committed groups may be more open to changing their behaviour 'radically', as the following respondent demonstrated:

> Well I think me and my wife, we're very open to things if they're made easy so any new scheme that comes along. (Focus group 3, Committed environmentalist)

The clear implication from these findings is that lifestyle groups require nuanced messages according to the level of behavioural change required. For the committed groups, this may involve more 'radical' suggestions of how to change lifestyle practices than for the less committed groups, who may need to have a constant reassurance that qualify of life will not be reduced by shifts in behaviour to help the environment.

The Intention-Behaviour Gap

> I like taking a really long shower in the morning and nothing is gonna stop me although I know I shouldn't in terms of water conservation. (Focus group 5, Occasional environmentalist)

The research presented in Chapter 8 and within this chapter has demonstrated a significant 'gap' between a desire to act in an environmentally-responsible manner and reported behaviour. Implicit within many responses to this issue (as demonstrated in the quotation above) was that people 'knew' their behaviour was 'wrong', but they persisted in any case. However, there is evidence from the focus groups that the discord between intentions and actions was mediated by the level of commitment required. Whilst the respondent quoted at the start of this section reflected on the relative small sacrifice involved in reducing the length of a shower, the following Committed environmentalist was more concerned with a more radical activity:

> Well like with water. I know if I've got a dirty washing up bowl I should empty it outside on the plants but it's cold and it's wet and it's the middle of the night so I don't, it goes down the drain instead. And I think it does worry me something because I think 'I should be doing this' (Focus group 3, Committed environmentalist).

The implication of this finding is that policy needs to set relative levels of 'expectation' in terms of behavioural commitment, such that intentions can more effectively be transferred (partially) into actions. Setting 'the bar' too high will exclude certain lifestyle groups, who will not feel able to undertake the behaviours being recommended. Once again, this will involves targeting and messages that are specifically nuanced regarding the expectations of behaviour change.

Identifying and acting to remove barriers   Building on the second and third concepts outlined previously, the more effective transformation of relative intentions into actions was clearly seen as being dependant on the removal of specific barriers to behaviour within both the quantitative and qualitative research outputs. Different groups identified alternative sets of barriers. These are detailed both in Chapters 7 and 8 and expanded upon earlier in this chapter. Some of these barriers are structural, relating to facilities provided for specific activities such as recycling. However, others are related to two key issues that have significant policy implications.

First, emphasised in different ways by each of the lifestyle groups, the *ascription of responsibility* for environmental problems was framed in varying terms. For the Committed environmentalists, responsibility was more an issue that lay with the individual. As this respondent stated:

> Well I think we should be able to [take action] regardless [of external factors]. (Focus group 3, Committed environmentalist)

However, although the quantitative analysis and focus groups showed that committed individuals were aware of their responsibilities, those in less committed groups framed responsibility in terms of organisational contexts, from the national to international level:

> Yeah, I feel the same way like what am I gonna do. America isn't even signing up to the Kyoto Agreement and you think it's such a big country and if it's not doing that and their petrol is so cheap, then what difference am I going to make? I care about the environment but don't see that I'll make a difference. (Focus group 8, Non-environmentalist)

The policy implication of these varying levels of ascribed responsibility is that alternative lifestyle groups require different messages to engender or reinforce levels of personal responsibility. In the case of less committed individuals, this must be seen within the context of 'trust' in 'official' information and policy-making, where levels of response efficacy are particularly low. Conversely, in the case of committed individuals messages need to emphasise the personal responsibilities that they have towards helping the environment.

Second, all of the lifestyle groups highlighted the role of what was termed '*big business*' in 'preventing' environmental action. Once again, emphasis was nuanced between the groups. Committed environmentalists often tended not to shop in supermarkets, but implicitly recognised the barriers encountered by those who felt they 'had to':

> I think a high percentage [explaining why people use supermarkets] is cost because as an example, the farmer's market, which I highly support, is much more expensive than anything you can buy in the supermarket. (Focus group 3, Committed environmentalist)

Supermarkets were seen as being convenient by many respondents and when asked whether more 'local' shops would change behavioural practice, a typical response was:

> It would still only make as convenient as the supermarkets but it wouldn't ever be more convenient so I don't think it'd change mine either as you'd always get more choice at the supermarket. (Focus group 5, Occasional environmentalist)

The clear implication of this finding is that for those who are less willing to adopt radically alternative consumption patterns, policy-makers need to work with supermarkets to promote more sustainable consumption practices, both in terms of their own purchasing and production mechanisms and also how consumers are engaged to purchase 'alternative' products. This is particularly challenging given the scepticism with which supermarkets were viewed by all groups. One means by which 'big business' might tackle this issue is by developing a series of brands that can be targeted at specific lifestyle groups. The research provided in the focus groups clearly demonstrates that brands can be utilised as powerful social marketing tools for influencing behaviour change. Two comments from participants emphasise this point:

> It's just that the brands that I buy, I know I'm going to get quality, I know that I'll get what I get for the price and I get stuck to it because of that. (Focus group 6, Non-environmentalist)

> Yeah or when you go to a certain store because you know if you walk in there you'll find a style you'll like because you know what their style is and things along those lines. (Focus group 6, Non-environmentalist)

Collective action   The low levels of response efficacy reported in the preceding sections by those less committed to environmental action is reflected both in the data from the questionnaire and focus groups. All focus groups argued that collective action was necessary to achieve the goals of sustainability, although they viewed their own specific roles within this very differently. Committed environmentalists, not surprisingly, saw themselves as 'awareness raisers' for collective action:

> Yeah, like word of mouth. Me telling that girl to put her rubbish in the bins provided probably made her more aware to do it in future. (Focus group 3, Committed environmentalist)

However, those less committed, whilst recognising the benefits of collective action, focused more on why it wasn't feasible:

> … it's everyone and we're just one person. So if we do something, then it doesn't actually do much unless we all do something. (Focus group 5, Occasional environmentalist)

Policy-makers therefore need to consider how those who are more committed to environmental action can be involved in raising awareness. Committed environmentalists can act as community champions – establishing norms for environmental action, by demonstrating levels of commitment that can be used to counter arguments of weak response efficacy. Successful schemes such as Global Action Plan's 'Small Change' and EcoTeams programmes are good examples of these community-led programmes. This links explicitly to the third concept previously identified, related to social marketing as a tool for changing behaviour and suggests that locally-based, socially contextual and community-led programmes of behaviour change can be effective and engender a clear sense of collective action and purse for individuals.

Incentives, not penalties   A final observation from this study is that incentives were highlighted by all focus groups as a way forward. Given the wide spectrum of environmentalism observed between the different lifestyle groups, incentive schemes would need to be targeted at particular clusters and should reflect the level of the behavioural shift required. However, it is likely that forms of incentives could play a significant role in bolstering community-based social marketing campaigns amongst specific lifestyle groups. This should be seen within the context of branding, which the Final Summary Report highlights as a major element in determining individual lifestyle choices.

Policy Implications: Conclusions

The policy implications provided in the preceding sections provide a series of key foci for the development of both strategic and specific policy measures. The major outcomes of the research for both ESRC and DEFRA relate to the critical role of understanding how environmental practices relate to everyday practice, the significance of segmenting individuals into lifestyle groups that can be used as the basis for a social marketing programme. This programme needs to involve the identification of barriers and motivations for action for each target segment. The policy implications derived above point to a series of barriers that need to be removed for specific lifestyle groups. These include notions of responsibility and response efficacy. However, the implications also support the notion outlined at the beginning of this chapter, which emphasised the more specific role of social marketing to both use local and socially contextual methods to encourage behaviour and using marketing and branding strategies to encourage behaviour. This will now be explored in the form of a concluding section of this chapter.

Conclusion: Changing Behaviour through Social Marketing?

This chapter has revealed a number of key strategic directions that policy can follow in order to make behaviour change policies more effective. The potential for using both a broadly defined and nuanced social marketing approach has been outlined in the preceding sections, emphasising a positive notion of behavioural shifts, nuanced for different lifestyle groups and centre on overcoming barriers to change. In many cases, policies for generating nuanced messages and removing key barriers are ones that do not need to be discussed within this book, but two remaining issues are evident from the policy implications which merit further research and development:

• the notion of using 'collective action' for change;
• the utility of 'marketing and branding techniques' for promoting change.

These two concepts focus on the related theme of 'social desirability' (Sadalla and Krull 1995), which focuses around ensuring that environmental action is embedded within the collective consciousness of everyday practices undertaken by individuals and groups, representing a series of behaviours that are aspirational rather than marginal. This implies, as was emphasised in Chapter 6, that environmental practices need to become embedded into everyday routines and habits, with consumption choices related to the environment made outside the home made within the same commercial context as other product choices. For example, there is great potential for environmentally-friendly goods to be subject to the same branding and marketing techniques that organisations use to promote most other products. Given the power of brands to persuade large number of consumer to purchase specific goods, it ought to be possible to apply such marketing strategies to individuals, a strategy that would incrementally enhance a collective consciousness.

There has been little academic research into the potential impact of using commercial marketing techniques to promote environmental practices. Accordingly, a number of key areas for research development can be identified. These need to focus around the use of key social marketing concepts, such as 'community champions', branding and the use of incentives to encourage behaviour change.

First, outside a commercial context, the role that Committed environmentalists might play in encouraging those with less commitment might play in developing community-based social marketing strategies is worthy of further study. Such an approach would rely on key social actors in a community promoting particular types of behaviour. This has been highlighted in terms of the role that 'Community Champions' can play in developing collective behaviour by charities such as Global Action Plan (GAP 1999). Crucially, our findings imply that the reinforcement of a sense of collective action at the local level is likely to tackle low levels of response efficacy highlighted in the research. Having ownership over a set of activities or policies can engender a greater sense of both self- and response-efficacy. Nonetheless, the choice of such champions would need to be undertaken with care, given that those in less committed lifestyle groups would react negatively to being encouraged by individuals with whom they do not share similar aspirations and values. However, the role of 'champions' needs to be seen within the context of developing two further broad areas of an environmental strategy that could reinforce any changes made at the community level.

The second of these broad areas relates to the role that branding can play in both promoting environmental action amongst key groups and, in tandem, enabling businesses (who have a negative perception amongst some lifestyle segments) to brand and market more sustainable forms of consumption to key market segments. Further research is needed to assess how environmental 'brands' could be used to convince specific lifestyle groups enough to change their behaviour. Our current research suggests that many of those in less committed groups have consigned themselves to using large supermarkets for their shopping, but simultaneously have little faith in the environmental credentials of these organisations. There is therefore scope to use a 'captive audience' to develop market-specific brands that could promote environmentally-friendly goods using the principles of marketing. This may involve considerable investment in relationship marketing techniques, which can be used to establish 'brand identities' for individuals. Crucially, for less committed groups who are the least interested in the environment, this may involve branding environmental goods according to key aspirational characteristics of such lifestyle groups. Accordingly, for some individuals, there might not be a need to mention the fact that a product was environmentally friendly. Such techniques would provide a means by which to target lifestyle groups who are currently out of range of existing environmental campaigns. Nonetheless, it may be that these 'voluntary' schemes may not be sufficient and so, in line with the development relationship marketing policies, a third dimensions to a social marketing approach can be envisaged.

Related to the notion of branding, a third and final social marketing approach would explore the role that using incentives to encourage lifestyle changes for particular groups could play in changing behaviour. The research reported in this text suggests that targeted incentives would encourage individuals to be more pro-environmental and, once again, could be linked to sustainable forms of consumption promoted by businesses. The National Consumer Council (NCC 2004) has already proposed the idea of an environmental 'credit card', which would be used to gain points on purchases in participating retail outlets, with 'double points' being awarded for products that are environmentally friendly. Such a scheme would enable a close relationship to be developed between the card promoter and consumer. In one sense, such a scheme sounds extremely attractive, given the power that supermarkets have in promoting the use of store loyalty cards and offering goods for 'extra points' to encourage sales of such products. Nonetheless, the persuasion of both credit card companies and retailers to participate would be problematic unless commercial benefits were evident. Indeed, the selection of which types of products to attract 'points' would be challenging for any regulatory body. Despite these reservations, further research is required to examine what level of incentives would be required, at which spatial scales and exactly what behaviours would need to be targeted at specific lifestyle groups.

These three areas for development are closely related in the move towards embedding sustainable lifestyles into everyday practices. The three environmental practices that were identified in Chapter 6 can be tackled within the context of such social marketing techniques. The use of branding and incentives evidently has the potential to impact on the purchasing behaviour of different lifestyle groups, whilst more locally-based community social marketing may be required to encourage habitual behaviours, such as energy and water conservation. Indeed, a combination of both community based social marketing and incentives could be used to encourage recycling behaviour. Such a combination of policies can offer a new agenda for encouraging behaviour change and presents significant opportunities for research into the role of social marketing as a tool for behaviour change behaviour. These all emphasise one central point that has been made in this and preceding chapter, namely that encouraging lifestyles that are commensurate with sustainable development will not be achieved by calls for radical shifts in behaviour or indeed broad-brush appeals to the population through campaigning. Rather, a focus on how people live their everyday lives, the lifestyle groups within which they are situated and the aspirations they have will present more effective ways to embed the notion of sustainable development into a society that is driven by priorities that currently do not place the environment at the top of the agenda.

Chapter 10

Sustainability, Citizens and Progress

Introduction: Sustainability and Society

This final chapter will reflect on the growing importance of social dimensions to sustainable development and the increasing attention being paid by policy-makers to 'human' solutions to environmental problems within the context of the empirical research reported in Chapters 6 to 9. The chapter begins by reviewing the critical shifts in our understanding of the relationship between sustainable development and society, before exploring the dominant themes to emerge from the research and the potential for further work in this area.

At the start of this book a framework for exploring sustainable development was outlined (Figure 2.1), highlighting various levels of understanding (principles, concepts, applications). Within the 'concepts' category, 'time', 'space' and 'capital' were presented as three lenses through which to view sustainability. From the perspective of this book, 'capital' has been critically important in framing a position on sustainable development. Yet as was seen in Chapters 2 and 3, the notion of capital has traditionally been framed in economic terms. However, this book has argued that since the early 1990s there has been the emergence of a new '*social* capital' for sustainable development in the UK, incorporating a growing agenda for considering the role of individuals in promoting sustainable development. Accordingly, the central message that emerged from Chapters 2, 3 and 4 related to the increasing role of individuals within the sustainability agenda. This change can be characterised by three dominant processes that will be examined in the following three sections. They relate to major changes in sustainable development at three scales: conceptual, strategic and political.

The Limits to Brundtland: From Global Government to Local Governance

Conceptually, what characterised *The Limits to Growth* report (Meadows et al. 1972) and subsequent responses during the 1970s was the focus on institutional, legislative and regulatory reform to address the conflict between economic growth and environmental protection. Initiatives such as the United Nations Environment Programme and the Man and the Biosphere project all placed this conflict at their core and sought to understand and create policies to ameliorate the discords between rapid economic expansion and an equally degrading ecosphere. The processes by which such understandings and policies could be formulated were framed within scientific (and many would say, technocentric) terms. Once understood, policy formulation was a matter of introducing regulation or legislation (although of course, this was voluntary at the global scale). Yet by 1983, the United Nations had recognised that such high-level engagement was proving ineffective in many cases, with degrading environmental processes still worsening in both the developing and developed world. The commissioning of Gro Harlem Brundtland to undertake a study of the nature of economic development and environmental degradation heralded a new approach to the study of human-environment relations. Brundtland was concerned with the processes that caused specific problems, with an emphasis on considering the role of local and indigenous individuals. In the final report of WCED (1987), Brundtland highlighted the role of individuals and previously unrecognised social groups in providing the key to solutions for sustainable development. She emphasised the need to derive localised, socially-contextual and specific policies for sustainable development embedded within cultural practices. At the heart of this shift from high-level decision making was a shift from 'government' to 'governance', in particular the formation of local plans for sustainable development. Accordingly, at the conceptual level, Brundtland's WCED (1987) report emphasised the key role that social actors at the local level had to play in creating and managing policies for sustainable development.

UK Strategy: From Environment to Society

Strategically, the UK reacted rapidly to Brundtland's (WCED 1987) report (DoE 1988). In the first UK Sustainable Development Strategy (DoE 1994), many of the principles highlighted by Brundtland were evident, but the strategic objectives related mainly to a definition of sustainable development that emphasised the role of existing environmental policy in managing environmental problems. However, the Labour Government's 1999 Strategy *A Better Quality of Life* (DETR 1999) re-focused UK strategic policy-making on sustainable development towards a social agenda, emphasising sustainable community development and the alleviation of poverty almost to the detriment of the environment. This balance has been re-adjusted in the most recent Strategy (DEFRA 2005), with the goals of sustainable development regarded as living within environmental limits and creating sustainable communities. These shifts are representative of the changing priorities of two governments, but they also represent a strategic acceptance that sustainable development involves the inclusion of wider social goals and therefore represents a strategic shift towards the conceptual basis for sustainable development that Brundtland has envisaged.

UK Policy: From Awareness to Understanding

Politically, whilst government has therefore recognised the value of society (and individuals) within the concept of sustainable development, the actual formation of policy towards specific problems has been slower to catch up with these conceptual and strategic shifts. The focus of this book has been on changing behaviours to create a more sustainable future. As highlighted in Chapter 4, the political approach to behaviour change has been grounded in the expectation that communicating the 'problem' of environmental degradation would be sufficient to make significant changes in peoples' behaviour. The approach characterised by almost all policy in the 1990s was the A-I-D-A model (Awareness–Information–Decision–Action). This linear model of behaviour change has been reflected in numerous policy instruments, such as the 'Going for Green' initiative. At its heart, A-I-D-A assumes that individuals are predisposed to undertaken environmental behaviours. They support the notion of helping the environment and, by definition, they will surely act accordingly if shown the scale of the problem and a practical solution? Indeed, such a model also assumes that this pro-environmental perspective is shared by the vast majority of the population.

Yet as detailed in Chapter 4, academic research has demonstrated the futility of national advertising and awareness campaigns. This is slowly being recognised in Whitehall, where DEFRA have been commissioning research on behaviour change for the past couple of years (Darnton 2004a; 2004b; Barr et al. 2006). There is now a tangible recognition in government that behaviour change is complex (DEFRA 2005) and specific, diverse and nuanced policies will be required to effect change. It will be some time before policy shifts to implement an alternative approach to behaviour change in government, but this book offers a starting point from which to base both a new research agenda and innovative political approach to changing behaviour for sustainable development. Accordingly, by way of setting these new agendas we now turn to examine the contribution that this book makes to both recent academic and political debates on environmental action.

A New Research Agenda: Geography, Psychology and Policy

Academic research into environmental action has emanated from a wide variety of disciplines and, as demonstrated in Chapter 4, has drawn assumptions from a series of epistemological and methodological understandings. However, the academic agendas on environmental action have been forged by researchers in two major disciplines, with their epistemological and methodological approaches varying significantly. Both of these agendas have been developed within the context of a dominant 'model' of behaviour change advocated by policy-makers.

As noted in the previous section, until very recently policy-makers have used the 'linear' model of behaviour change, implying that behavioural shifts are the result of decisions make by individuals, on the basis of information provided about environmental issues. Chapter 4 provided a number of examples of this type of policy-making, grounded in key assumptions concerning both the transfer of knowledge from the state to the individual and the efficacy with which such information would be translated into behavioural commitments by a majority of the population.

The Geographical Approach

As Chapter 4 noted, a strong criticism of this approach to behaviour change has been evident from the geographical scholarship on environmental action. Such criticisms have focused around the weaknesses of the logic implicit in the 'linear' model and the critique of 'rationalistic' models that assume a direct behavioural response from positive attitudes towards behaviour change (Owens 2000). In particular, questions have been raised concerning the extent to which individuals are willing to accept narratives of environmental change prescribed by the state (Eden 1993) within a 'top-down' policy framework. This 'expertisation' of environmental issues has, according to Eden (1993), led to uncertainty and disengagement amongst lay publics, who are unable to sense and verify 'scientifically' defined environmental problems.

This emerging 'distance' between science and society (or 'expert' and 'lay') has led geographers such as Burgess et al. (1998) and Eden (1998) to advocate a 'deliberative' approach to environmental action. This 'civic' model (Owens 2000) seeks to engender greater levels of environmental responsibility through an embedded approach to environmental action, whereby citizens define environmental problems and create strategies for their amelioration at the community level. Along with the critique of 'linear' decision-making, geographers have provided a radically alternative approach to encouraging environmental action, based on contextualised understandings of behaviour and grounded in a civic and deliberative approach that places inclusivity and consensus at its centre. Such an approach appears compelling in the first instance. However, a central assumption made relates to the power and effectiveness of civic and deliberative approaches to change behaviour. Such a framework for behaviour change relies upon high levels of political interest and activity from the majority in society. However, as the research for this book has illustrated, those individuals who are the least likely to help the environment are also those who are the most politically apathetic and therefore less likely to engage in any novel deliberative process. Accordingly, whilst the civic approach advocated by geographers sounds compelling, it fails to fully realise many fundamental political and social realities.

A Social-Psychological Alternative?

However, there is a further rationale for questioning the logic of the dominant geographical discourse regarding environmental action. As noted in Chapter 4, a central point of contention between what Owens (2000) terms the 'mental models' approach and a civic/deliberative model is one defined less by epistemology and more by methodology (Barr and Gilg 2006). Burgess et al. (1998) and Eden (1993; 1998) have critiqued the use of quantitative survey techniques and statistical methods to examine environmental behaviour. Part of this criticism emanates from the stark differences in the methods applied to the study of environmental action by geographers on the one hand and social psychologists on the other. As noted in Chapter 4, the social-psychological tradition of studying environmental action has a far longer history and utilises a theoretical approach based on wider social-psychological concepts. The work of social (and of course environmental) psychologists has been critiqued by geographers for being 'rationalistic' and still based on a linear model of behaviour change. Such accusations are largely unfounded and the complexity of psychological research into behaviour change demonstrates the ability of psychology to make a significant contribution to the academic debate. Two highlights of the social-psychological approach can be selected to demonstrate the effectiveness of this approach. First, the use of existing social theories of behaviour change (such as the Theory of Planned Behaviour; Ajzen 1991) enable this approach to compare behavioural shifts across a range of domains and provide a transparent theoretical grounding for research. Second, the use of existing and reliable measures for specific attitudes enables the testing of alternative approaches to behavioural change. Taken together, these advantages illustrate that social psychology can provide a large contribution to the debate on environmental action. However, this book has moved the debate forward beyond these disciplinary boundaries to illustrate the utility of a new theoretical and methodological approach to studying environmental action. This is commensurate with calls by geographers such as Wilson (1997) and Burton (2004) who have argued for a wider use of social-psychologically informed understandings of behaviour change within geography. Accordingly, the next two sections provide the basis for forming a new theoretical and policy agenda for studying environmental action within geography, relying heavily on understandings drawn from environmental psychology and driven by a desire to reconnect geography with emerging policy agendas.

A Theoretical Framework of Environmental Action

The research in this book supports the notion that using the principles of social-psychology can yield a wide variety of data that can directly inform policy. However, this book sought not to repeat the vast array of studies undertaken by social-psychologists in this field, but to demonstrate to geographers and other researchers the value of drawing on a wider range of variables and concepts than have previously been utilised by social-psychologists. This approach is grounded in the use of a theoretical framework of environmental action (Figure 4.3) that has been developed by the author over the past five years. This book has demonstrated the utility of this framework through the use of three distinctive methodologies, which have enabled the identification of different types of environmental behaviour, alternative lifestyle groups and the influence of each set of factors in the framework on each of these groups. Indeed, the framework offers a major contribution to the debate concerning environmental action and behaviour change. First, the framework presents a flexible approach to understanding environmental action. To this end, it is assumed that alternative environmental behaviours will be defined (as seen in Chapter 6) and that such behaviours will be influenced by a combination of social and environmental values, situational characteristics and psychological factors. Second, the framework acknowledges that there is an implicit conflict between behavioural intention and reported activity, a central element of Fishbein and Ajzen's (1975) Theory of Reasoned Action. The so-called 'value-action' gap has attracted widespread attention in both the media and academic circles, with attention focussing on the disparities between stated intention to help the environment but a weak environmental commitment. Third, the framework offers significant opportunities for policy-makers. The diagrams produced in Chapter 8 clearly demonstrate the critical variables involved in determining environmental action for alternative lifestyle groups, which were also reflected in the focus group work reported in Chapter 9. Accordingly, the framework presents a significant contribution to the debate concerning the structure and understanding of environmental action and contributes to the theoretical debate summarised by Owens (2000) concerning understandings of environmental consciousness.

An Agenda for Political Change: Three Key Principles

The research reported in Chapters 6, 7, 8 and 9 point to three key principles that academics and policy-makers need to appreciate when studying environmental action and to this end the central theme relates to the methodological processes undertaken during the research and policy formation stages. These relate to the ways in which environmental practices are framed, the different lifestyle groups in society that perform these practices and the potential changes that need to be made in political terms to shift behavioural patterns.

Environmental and Everyday Practices

The first key issue that academics and policy-makers need to examine relates to the ways in which environmental practices are conceptualised and promoted. The dominant disciplinary and practical boundaries placed on environmental action have been framed around environmental 'problems'. From an academic perspective, the framing of environmental practices has been contextualised within disciplinary boundaries that have reflected traditional academic discourses. From a political viewpoint, environmental practices have been framed around the fluctuating urgency with which a specific environmental issue rises to the top of the political agenda.

Neither of these approaches offers an appropriate response and understanding of environmental practices. The research reported in this book has demonstrated the critical importance of removing these disciplinary and practical boundaries to enable understandings of environmental practices that are placed within the context of everyday practices. These are framed around the decisions taken when purchasing goods and services, the everyday habits within the home and the distinctive act of recycling, performed in and around the home. Environmental actions are not undertaken in behavioural isolation from other behaviours; they are reflective of the everyday practices of individuals, being undertaken in both consumptive and habitual contexts, alongside a myriad of other activities. Accordingly, both researchers and policy-makers need to shift their attention towards understanding 'practices not 'problems', moving their research and political agendas towards a focus on lifestyles.

Sustainable Lifestyles

This focus on lifestyles needs to take account of these differing practices. However, if it is accepted that our attention needs to shift towards practice, not problems, then a second concession that researchers and policy-makers need to make relates to the nature of sustainable lifestyles. As was seen in Chapter 4, previous studies of environmental behaviour and policies for changing behaviour have assumed that the population is relatively similar, with pro-environmental values, but lacking in the ability to act out these values. The research reported in this book challenges this notion of behaviour change, arguing that people's environmental practices are reflective of their everyday practices and therefore reflect their lifestyle. As Chapter 7 highlighted, market segmentation research and application since the 1960s has stressed the importance of defining and using alternative market segments to promote commercial goods. The research in this book has illustrated the utility of such a methodology for examining the importance of lifestyle groups insofar as they relate to environmental practices. In the research reported here, four types of individual were identified who reflected a series of lifestyle types, reflected in differences according to social and environmental values, demographic characteristics and psychological factors. Although the quantitative research was undertaken in one area of the UK in 2002, these spatial and temporal limitations do not preclude some broad generalisations to be made concerning the nature of environmental lifestyles in Britain today.

The Committed Environmentalists identified in Chapter 7 represented a highly motivated group of individuals, identifying themselves by their commitment to the environment. They undertook a wide range of environmental practices and were demographically distinctive. Mainstream Environmentalists were similar in many ways to the committed group, although they were less likely to undertake activities such as composting. However, greater differences emerged between this group and both the Occasional and Non-environmentalists, representing a far weaker commitment to the environment. Indeed, the Non-environmentalists comprised a socially-distinctive group and may well represent a wider social profile recognised in society more generally. Although these profiles are unique to this study, they do illustrate the importance of examining the potential differences between lifestyle groups, which leads to a third and politically crucial point.

Lifestyles and Behaviour: The Value-Action Gap

Chapter 8 demonstrated the importance of understanding the different influences that acted to drive behaviour for the four different lifestyle groups identified. Crucially, this chapter highlighted the differences between intended behaviour and reported action. Indeed, for each lifestyle group, very different factors were seen to govern alternative environmental practices. This was also highlighted in Chapter 9, where the focus group research determined the different factors acting on each lifestyle segment.

Chapter 9 also demonstrated the range of policy implications that could be derived from an understanding of these factors and highlighted the potential role for social marketing techniques to contribute towards behavioural changes. The notions of promoting 'collective action' based on the removal of key barriers to change alongside and 'branding' for sustainability were introduced as ways of exploiting the potential for developing a lifestyles approach towards environmental practice. Such a shift would be fundamental; it would involve a reconceptualisation of what it means to be green, a rewriting of the political textbook for behaviour change. Most of all, it would mean being positive about change and would necessitate the use of commercial marketing techniques for a problem which has been traditionally viewed as an issue for the state to resolve. Yet there are signs that a broadly-defined social marketing approach is being adopted by government and business, with the incentives being provided by some supermarkets for re-using shopping bags and the emergence of eco-labelling as a device for enabling consumers to make more informed decisions about their purchases. Nonetheless, we are still a long way from a market-based system for creating incentives for environmental action and more research is needed to effectively discern the role of social marketing and branding as a means by which to promote behaviour change.

A New Geography of Environmental Practice?

The three themes set out in the previous section provide a compelling agenda for researchers in the coming years. Although geography still reflects the impact of the 'cultural turn', the emerging political agendas of the twenty-first century necessitate a shift towards research approaches that speak to the wide range of national and local policy-makers who are seeking greater understandings of how to effect behaviour change. The debate concerning 'relevance' in geography has been ongoing for a number of years and this book adds its weight to arguments that have advocated a more pro-active approach to engaging with policy-makers at all levels. This inevitably involves moving beyond criticisms of current political discourses and presenting theoretically informed and rigorous alternatives. Although the civic and deliberative model of behaviour change advocated by many geographers is compelling, it does not reflect the political realities of an apathetic and uninterested society. The approach advocated within this book attempts to bridge the gap between a naïve political approach (the 'linear' model of behaviour change) and existing geographical research by utilising social-psychological work to present a theoretically-informed framework of environmental action. This is operationalised by three inter-linked methodologies, uncovering the nature of environmental practice, the complexity of lifestyle groups and the factors that influence each of these behaviours for each group. These methodological developments permit a new approach to changing behaviours, through the use of market segmentation.

Future Theoretical Work

The 'end point' of the current research was described in Chapter 9. Although the utility of using segmentation techniques to identify both environmental practices and sustainable lifestyle groups was successful, the focus groups revealed specific trends that require further investigation. These relate to three emerging research agendas.

First, as outlined in Chapter 9, central to any further investigation will be an exploration of the connections between policy formation and implementation between the state and the commercial sector. The potential role that could be played by the commercial sector and the impact of social marketing for changing behaviour need to be studied further. Theoretically, work needs to focus on the role of commercial marketing and branding techniques in presenting strategies of behaviour change. Given the reliance individuals place on supermarkets and other 'big businesses' and the role of brands in forming 'brand identities', novel social and relationship marketing techniques could provide a means by which to encourage individuals to make more sustainable decisions. Such research will need to engage with theories of marketing and will require a wider understanding of the lifestyle groups highlighted in this book. Accordingly, a second wider development is required.

This second development relates to the role that environmental actions in and around the home have to behaviours at work and in leisure contexts. Most research has focused around the examination of environmental action in specific contexts (mostly in the home) and future research will need to examine the role of alternative environments in formulating environmental attitudes and behaviour. In particular, research will need to address the role of place in determining behaviour and question whether some 'places' become spaces where environmental behaviours are more or less likely to be performed. The notion of creating alternative behavioural spaces and inverting behaviour within one space (such as a leisure space) needs to be examined in the context of evidence that indicates individuals perform environmental practices in specific contexts. For example, Committed environmentalists may demonstrate a range of pro-environmental behaviours whilst in the home or at the supermarket, but may 'invert' or even negate this activity by choosing to fly to Spain for a holiday. Such a research agenda also needs to take a greater holistic perspective on 'lifestyles', incorporating both environmental and non-environmental behaviours. Most crucially, the notion of environmental practices needs to be investigated to understand the links between overtly environmental practices (such as green forms of consumption) and wider patterns of consumption, such as air and road travel.

Finally, research is needed that focuses on the long-term adoption of environmental practices and the formation of environmental lifestyles. As was seen during the empirical parts of this book, lifestyle groups each had different levels of behavioural commitment to the three environmental practices identified. Longitudinal research would enable a greater understanding of how environmental lifestyles and practices are developed through time and which groups are more likely to be 'early adopters' of specific activities and what barriers are evident when seeking to adopt new practices. Taken together, these developments will develop both the theoretical framework of environmental behaviour and widen the disciplinary scope of environmental behaviour research.

Limitations and Qualifications

The research reported in this book, as highlighted in Chapter 2, is driven from a geographical perspective and utilises classical social scientific assumptions to study environmental action. It is recognised that a range of perspectives and approaches exist towards studying environmental action. The approach advocated here is driven by two motivations. First, a belief in the scientific method (as evident in Chapters 6 to 8) as the only reliable and verifiable approach to studying social phenomena. In recent years, geographers have come to critique the assumptions underlying the scientific and empirical approach to research, a concern which has thankfully less troubled other social science disciplines. Accordingly, the scientific and empirical assumptions made by social-psychologists have begun to attract researchers from geography who have sought to redress the balance away from culturally-informed approaches. Second, the approach advocated in this book derives from a tangible sense that geographers should be engaged with policy and policy-makers. This involves both criticism, but also engagement and constructive development.

Bearing these qualifications in mind, the major limitations underlying the research reported in this book pertain to the scale of the research, both in time and space. Temporally, the research on which this book is based was a 'snapshot' of environmental action at one point in time. The lifestyle groups on which this research was based are therefore subject to the criticism that these may not represent lifestyle groups, but rather cohorts, which are representative of lifecycles rather than lifestyles. Spatially, the research was undertaken in Devon, a relatively homogeneous rural county in South West England. Although urban areas were surveyed, the diversity that would be present in larger cities and conurbations was not measured in this work. In particular, the impact on lifestyle groups of ethnicity and a wider range of incomes could not be assessed. Accordingly, the data need to be regarded as both spatially contextual and temporally distinct.

Conclusion: Reframing our Green Dilemmas

At the start of this book, three green dilemmas were outlined, relating to the tensions between conservation and growth, individual and society and the rupturing of scale in time and space. Throughout this book these dilemmas have been explored using data from empirical research, progressively problematising the dominant political perspectives which have argued that behaviour change is a matter of (simple) awareness-raising. Through analyses of both quantitative and qualitative data, it has been demonstrated that promoting environmental action is as complex a social behaviour to develop as any other. Accordingly, the findings suggest that some in society are predisposed to environmental conservation, have an altruistic perspective and clearly envision their behaviours in terms of the potential impacts globally and for future generations. Others evidently take an alternative standpoint. It is to these individuals that political focus needs to shift and this book has provided one possible approach to overcoming the green dilemmas which many in society perceive when contemplating environmental action. Critical to this approach will be an acknowledgement that such green dilemmas will always exist and that for the majority in society, there is unlikely to be a major shift towards a conservation-ethic or indeed a Damascene conversion towards altruistic values. Policy makers therefore need to work within the framework of capitalist, growth-orientated values which the majority of the population hold. This will mean working with the tools of capitalism to effect changes. Such changes are possible. The growth of the Fair Trade movement illustrates the potential to utilise effective marketing and branding to change consumer behaviour. So, the future does not lie in a reductionist or apocalyptic view of global change, simply because few will listen to this message. The message that will have resonance is that which speaks to individuals in the language they are most receptive to. Accordingly, perhaps the greatest dilemma is not for individual citizens, but rather for politicians, who urgently need to recognise that their power to influence behaviour change does not lie in publicly-driven campaigns but in a far more potent agent for change – the consumer society.

Bibliography

Adams, W.M. (2001), *Green Development*, 2nd edn (London: Routledge).

Ainsworth, P. (2004), 'Leading by Example: The government needs to do much more', *Eg Magazine* 10:1, 1–2.

Ajzen, I. (1991), 'The Theory of Planned Behavior', *Organizational Behavior and Human Decision Processes* 50, 179–211.

Anderson, W.T. Jr and Cunningham, W.H. (1972), 'The Socially Conscious Consumer', *Journal of Marketing* 36, 23–31.

Arbutnot, J. (1977), 'The Roles of Attitudinal and Personality Variables in the Prediction of Environmental Behavior and Knowledge', *Environment and Behavior* 9:2, 217–32.

Baalderjahn, U. (1988), 'Personality Variables and Attitudes as Predictors of Ecologically Responsible Consumption Patterns', *Journal of Business Research* 17, 51–6.

Baldassare, M. and Katz, C. (1992), 'The Personal Threat of Environmental Problems as Predictor of Environmental Practices', *Environment and Behavior* 24:5, 602–16.

Ball, R. and Lawson, S.M. (1990), 'Public Attitudes towards Glass Recycling in Scotland', *Waste Management and Research* 8, 177–82.

Barnes, P.M. and Barnes, I.G. (1999), *Environmental Policy in the European Union* (Cheltenham: Edward Elgar).

Barr, S. (2002), *Household Waste in Social Perspective* (Aldershot: Ashgate).

Barr, S. (2003), 'Strategies for Sustainability: Citizens and responsible environmental behaviour', *Area* 35:3, 227–40.

Barr, S. (2004), 'Are We All Environmentalists Now? Rhetoric and Reality in Environmental Decision Making', *Geoforum* 35:2, 231–49.

Barr, S. and Gilg, A.W. (2003), *Environmental Action and Around the Home*, Final report for the Economic and Social Research Council (Swindon: ESRC).

Barr, S. and Gilg, A.W. (2006), 'Sustainable Lifestyles: Framing environmental action in and around the home', *Geoforum* 37:6, 906–20.

Barr, S. and Gilg, A.W. (forthcoming), 'A Conceptual Framework of Environmental Behavior', *Geografiska Annaler B*.

Barr, S., Gilg, A.W. and Ford, N.J. (2001), 'A Conceptual Framework for Understanding and Analysing Attitudes towards Household Waste Management', *Environment and Planning A* 33, 2025–48.

Barr, S., Gilg, A.W. and Ford, N.J. (2005a), 'The Household Energy Gap: The divide between habitual and purchase-related conservation behaviours', *Energy Policy* 33:11, 1425–44.

Barr, S., Gilg, A.W. and Ford, N.J. (2005b), 'Defining the Multi-dimensional Aspects of Household Waste Management: A study of reported behaviour in Devon', *Resources, Conservation and Recycling* 45:2, 172–92.

Barr, S., Gilg, A.W. and Shaw, .G (2006), 'Technical Paper: Providing the supporting analysis, methodological approaches and emerging findings' (unpublished technical paper).

Barr, S., Gilg, A.W. and Shaw, G. (2007a), 'Promoting Sustainable Lifestyles: A social marketing approach' (Unpublished Final Summary Report).

Barr, S., Gilg, A.W. and Shaw, G. (2007b), *Targeting Specific Lifestyle Groups: A practical guide* (London, DEFRA).

Basiago, A.D. (1995), 'Methods of Defining Sustainability', *Sustainable Development* 3, 109–19.

Beck, U. (1992), *Risk Society: Towards a new modernity* (London: Sage).

Beck, U. (1999), *World Risk Society* (Cambridge: Polity Press).

Berger, I.E. (1997), 'The Demographics of Recycling and the Structure of Environmental Behavior', *Environment and Behavior* 29:4, 515–31.

Berk, R.A., Cooley, T.F., La Civita, C.J., Parker, S., Sredi, K. and Brewer, M. (1980), 'Reducing Consumption in Periods of Acute Scarcity: The case of water', *Social Science Research* 9, 99–120.

Berk, R.A., Schulman, D., McKeever, M. and Freeman, H.E. (1993), 'Measuring the Impact of Water Conservation Campaigns in California', *Climatic Change* 24, 233–48.

Berkowitz, L. and Lutterman, K.G. (1968), 'The traditional Socially Responsible Personality', *Public Opinion Quarterly* 32, 169–85.

Black, J.S., Sterm P. and Elworth, J.T. (1985), 'Personal and Contextual Influences on Household Energy Adaptions', *Journal of Applied Psychology* 70:1, 3–21.

Blake, J. (1999), 'Overcoming the "Value–Action" Gap in Environmental Policy', *Local Environment* 4, 257–78.

Boldero, J. (1995), 'The Prediction of Household Recycling of Newspapers: The role of attitudes, intentions, and situational factors', *Journal of Applied Social Psychology* 25, 440–62.

Bourdieu, P. (1984), *Distinction: A social critique of the judgment of taste* (R. Nice, trans.) (Cambridge, MA: Harvard University Press).

Brandon, G. and Lewis, A. (1999), 'Reducing Household Energy Consumption: A qualitative and quantitative field study', *Journal of Environmental Psychology* 19, 75–85.

Bratt, C. (1999), 'Consumers Environmental Behavior: Generalized, sector-based or compensatory?', *Environment and Behavior* 31:1, 28–44.

Bulkeley, H. (1999), 'Valuing the Global Environment: Publics, politics and participation' (unpublished PhD thesis, University of Cambridge)

Bulkeley, H. (2001), 'Governing Climate Change: The politics of risk society?', *Transactions of the Institute of British Geographers* 26:4. 430–47.

Burgess, J., Harrison, C.M. and Filius, P. (1998), 'Environmental Communication and the Cultural Politics of Environmental Citizenship', *Environment and Planning A* 30, 1445–60.

Burningham, K. and O'Brien, M. (1994), 'Global Environmental Values and Local Contexts of Action', *Sociology* 22:4, 913–32.

Burton, R.J.F. (2004), 'Reconceptualising the "Behavioural Approach" in Agricultural Studies: A socio-psychological perspective', *Journal of Rural Studies* 20, 359–71.

Burton, R.J.F., Wilson, G. (2006), 'Injecting Social Psychology Theory into Conceptualisation of Agricultural Agency: Towards a post-productivist self-identity', *Journal of Rural Studies* 22, 95–115.

Cabinet Office (2004), *Personal Responsibility and Changing Behaviour* (London: Cabinet Office).

Cairncross, F. (1992), UNCED, 'Environmentalism and Beyond', *Colombia Journal of World Business* 27, 12–17.

Cameron, L.D., Brown, P.M. and Chapman, J.G. (1998), 'Social Values and Decisions to take Proenvironmental Action', *Journal of Applied Social Psychology* 28:8, 675–97.

Canter, D. and Donald, I. (1987), 'Environmental Psychology in the United Kingdom', in D. Stokols and I. Atman (eds), *Handbook of Environmental Psychology* (New York: Wiley), 1281–310.

Carson, R. (1962), *Silent Spring* (Boston: Mifflin).

Chan, R.Y.K. (1998), 'Mass Communication and Proenvironmental Behaviour: Waste recycling in Hong Kong', *Journal of Environmental Management* 52, 317–25.

Chan, R.Y.K. (2001), 'Determinants of Chinese Consumers' Green Purchase Behavior', *Psychology and Marketing* 18:4, 389–413.

Cloke, P., Crang, M. and Goodwin, M. (eds) (2005), *Introducing Human Geographies*, 2nd edn (London: Edward Arnold).

Coggins, P.C. (1994), 'Who is the Recycler?', *Journal of Waste Management and Resource Recovery* 1, 69–75.

Collins, A.J. (2004), 'Can We Learn to Live Differently? Lessons from Going for Green: A case study of Merthyr Tydfil (South Wales)', *International Journal of Consumer Studies* 28:2, 202–11.

Commission of the European Communities (1992), *Towards Sustainability: The Fifth Environmental Action Programme* (Luxembourg: CEC).

Commission of the European Communities (1997), *Towards Sustainability: the European Commission's progress report and action plan on the fifth programme of policy and action in relation to the environment and sustainable development* (Luxembourg: CEC).

Commission of the European Communities (2002), *The Sixth Environmental Action Programme* (Luxembourg: CEC).

Connelly, J. and Smith, G. (2003), *Politics and the Environment*, 2nd edn (London: Routledge).

Cooper, J.R.T. and Smyth, A. (2001), 'Contemporary Lifestyles and the Implications for Sustainable Development Policy: Lessons from the UK's most car dependent city, Belfast', *Cities* 18:2, 103–13.

Corraliza, J.A. and Berenguer, J. (2000), 'Environmental Values, Beliefs and Actions: A situational approach', *Environment and Behavior* 32:6, 832–48.

Corral-Verdugo, V. (1997), 'Dual "Realities" of Conservation Behaviour: Self reports Vs observations of re-use and recycling behaviour', *Journal of Environmental Psychology* 17, 135–46.

Corral-Verdugo, V., Bernache, G., Encinas, L. and Garibaldi, L. (1994–95), 'A Comparison of Two Measures of Reuse and Recycling Behaviour: Self-report and material culture', *Journal of Environmental Systems* 23:4, 313–27.

Costanzo, M., Archer, D., Aronson, E. and Pettigrew, T. (1986), 'Energy Conservation Behavior: The difficult path from information to action', *American Psychologist* 41:5, 521–28.

Couch, C. and Dennemann, A. (2000), 'Urban Regeneration and Sustainable Development in Britain: The example of the Liverpool Ropeworks Partnership', *Cities* 17:2, 137–47.

Counsell, D. (1999), 'Attitudes to Sustainable Development in Planning: policy integration, participation and Local Agenda 21, a case study if the Hertfordshire Structure Plan', *Local Environment* 4: 1, 21–32.

Cowell, R. (1997), 'Stretching the Limits: Environmental compensation, habitat creation and sustainable development', *Transactions of the Institute of British Geographers* 22:3, 292–306.

Darnton, A. (2004a), *Driving Public Behaviours for Sustainable Lifestyles: Report 1 of Desk Research commissioned by COI on behalf of DEFRA* (Andrew Darnton Research and Analysis).

Darnton, A. (2004b), *Driving Public Behaviours for Sustainable Lifestyles: Report 2 of Desk Research commissioned by COI on behalf of DEFRA* (Andrew Darnton Research and Analysis).

Darnton, A.R. and Sharp, V. (2006a), *Segmenting for Sustainability Report 1: Commentary* (Andrew Darnton Research and Analysis for the Social Marketing Practice).

Darnton, A.R. and Sharp, V. (2006b), *Segmenting for Sustainability Report 2: Commentary* (Andrew Darnton Research and Analysis for the Social Marketing Practice).

Davies, A.R. (2002), 'Power, Politics and Networks: Shaping partnerships for sustainable communities', *Area* 34:2, 190–203.

De Oliver, M. (1999), 'Attitudes and Inaction: A case study of the manifest demographics of urban water conservation', *Environment and Behavior* 31:3, 372–94.

De Young, R. (1985–86), 'Encouraging Environmentally Appropriate Behaviour: The role of intrinsic motivation', *Journal of Environmental Systems* 15:4, 281–92.

De Young, R. (1986), 'Some Psychological Aspects of Recycling', *Environment and Behavior* 18:4, 435–49.

De Young, R. (1988–89), 'Exploring the Difference between Recyclers and Non-recyclers: The role of information', *Journal of Environmental Systems* 18:4, 341–51.

De Young, R. (1990), 'Recycling as Appropriate Behaviour: A review of survey data from selected recycling education programmes in Michigan', *Resources, Conservation, Recycling* 3, 253–66.

De Young, R. (1996), 'Some Psychological Aspects of Reduced Consumption Behavior: The role of intrinsic motivation and competence motivation', *Environment and Behavior* 28:3, 358–409.

De Young, R. and Kaplan, S. (1985–85), 'Conservation Behaviour and the Structure of Satisfactions', *Journal of Environmental Systems* 15:3, 233–42.

De Young, R. and Robinson, J. (1984), 'Some Perspectives on Managing Water Demand: Public and expert views Canadian', *Journal of Water Resources* 8, 9–18.

Demos/Green Alliance (2003), *Carrots, Sticks and Sermons: Influencing public behaviour for environmental goals* (London: Demos).

Department for the Environment, Transport and the Regions (DETR) (1998), *Opportunities for Change: A consultation paper for a new strategy for sustainable development* (London: DETR).

Department for the Environment, Transport and the Regions (DETR) (1999a), *Limiting Landfill: A consultation paper on the European Union's Landfill Directive* (London: DETR).

Department for the Environment, Transport and the Regions (DETR) (1999b), *A Better Quality of Life: A strategy for Sustainable Development for the UK* (London: The Stationary Office).

Department for the Environment, Transport and the Regions (DETR) (1999c), *Quality of Life Counts* (London: DETR and Government Statistical Service).

Department for the Environment, Transport and the Regions (DETR) (2000a), *Waste Strategy 2000* (London, The Stationary Office).

Department for the Environment, Transport and the Regions (DETR) (2000b), *Are You Doing Your Bit? Campaign* (London: DETR).

Department for Transport (DfT) (2005), *Transport Statistics for Great Britain* (London: The Stationary Office).

Department of the Environment (DoE) (1988), *Our Common Future: A perspective by the United Kingdom on the Report of the World Commission on Environment and Development* (London: HMSO).

Department of the Environment (DoE) (1990), *This Common Inheritance: Britain's Environmental Strategy* (London: HMSO).

Department of the Environment (DoE) (1994), *Sustainable Development: The UK Strategy* (London: HMSO).

Department of the Environment, Food and Rural Affairs (DEFRA) (2004a), *Quality of Life Counts: Indicators for a strategy for sustainable development for the United Kingdom* (London: DEFRA).

Department of the Environment, Food and Rural Affairs (DEFRA) (2004b), *Taking It On: Developing UK sustainable development strategy together* (London: DEFRA).

Department of the Environment, Food and Rural Affairs (DEFRA) (2005), *Securing the Future: UK Government sustainable development strategy* (London: The Stationary Office).

Department of the Environment, Food and Rural Affairs (DEFRA) (2006), *Sustainable Development Indicators in Your Pocket* (London: DEFRA and National Statistics).

Derksen, L. and Gartell, J. (1993), 'The Social Context of Recycling', *American Sociological Review* 58, 434–42.

Dillman, D.A. (1978), *Mail and Telephone Surveys: The total design method* (Chichester: Wiley).

Dillman, D.A., Rosa, E.A. and Dillman, J.J. (1983), 'Lifestyle and Home Energy Conservation in the United States: The poor accept lifestyle cutbacks while the wealthy invest in conservation', *Journal of Economic Psychology* 3, 299–315.

Dunlap, R.E. (1975), 'The Impact of Political Orientation on Environmental Attitudes and Actions', *Environment and Behavior* 7: 4, 428–53.

Dunlap R.E. and Van Liere, K.D. (1978), 'The New Environmental Paradigm', *Journal of Environmental Education* 9, 10–19.

Dunlap, R.E., Van Liere, K.D., Mertig, A.G. and Jones, R.E. (2000), 'Measuring Endorsement of the New Ecological Paradigm: A revised NEP scale', *Journal of Social Issues* 56, 425–42.

Eagly, A. (1987), *Sex Differences in Social Behavior: A social role interpretation* (Hillsdale, NJ: Erlbaum).

Eckersley, R. (1991), *Environmentalism and Political Theory: Towards an ecocentric approach* (London: UCL Press).

Eden, S. (1993), 'Individual Environmental Responsibility and its Role in Public Environmentalism', *Environment and Planning A* 25, 1743–58.

Eden, S. (1996), 'Public Participation in Environmental Policy: Considering scientific, counter-scientific and non-scientific contributions', *Public Understanding of Science* 5, 183–204.

Ekins, P. and Max-Neef, M. (1992), *Real-life Economics: Understanding wealth creation* (London: Routledge).

Etzioni, A. (1993), *The Spirit of Community: Rights, responsibilities, and the Communitarian agenda* (London: Fontana/HarperCollins).

Etzioni, A. (1995a), 'Introduction', in A. Etzioni (ed.), *New Communitarian Thinking: Persons, virtues, institutions, and communities* (London: University of Virginia Press), 1–15.

Evans, B. (1995), 'Local Environmental Policy and Local Agenda 21', *Area* 27:2, 163–4.

Evans, B. (2003), 'We Need to Rescue Sustainable Development from the Cost Rhetoric', *Eg Magazine* 9:8, 1–2.

Exeter City Council (1996), *A Local Agenda 21 for Exeter* (Exeter: Exeter City Council).

Fishbein, M. and Ajzen, I. (1975), *Belief, Attitude, Intention and Behavior: An introduction to theory and research* (Reading, MA: Addison-Wesley)

Franks, T.R. (1996), 'Managing Sustainable Development: Definitions, paradigms and dimensions', *Sustainable Development* 4:2, 53–60.

Freeman, C., Littlewood, S. and Whitney, D. (1996), 'Local Government and the Emerging Models of Participation in the Local Agenda 21 Process', *Journal of Environmental Planning and Management* 39:1, 65–78.

Gaskell, G. (1983), 'Consumer Energy Research: Progress and prospects', *Journal of Economic Psychology* 3, 185–91.

Gibbs, D., Longhurst, J. and Braithwaite, C. (1996), 'Moving towards Sustainable Development? Integrating Economic Development and the Environment in Local Authorities', *Journal of Environmental Planning and Management* 39:3, 317–32.

Gibbs, D.C., Longhurst, J. and Braithwaite, C. (1998), 'Struggling with Sustainability: Weak and strong interpretations of sustainable development within local authority policy', *Environment and Planning A* 30:8, 1351–65.

Giddings, B. Hopwood, B. and O'Brien, G. (2002), 'Environment, Economy and Society: Fitting them together into sustainable development', *Sustainable Development* 10, 187–96.

Gilg, A.W. (1996), *Countryside Planning* (London: Routledge).

Gilg, A.W. (2005), *Planning in Britain: Understanding and evaluating the post-war system* (London: Sage).

Gilg, A.W. and Barr, S. (2005a), 'Green Consumption or Sustainable Lifestyles? Identifying the Sustainable Consumer', *Futures* 37:6, 481–504.

Gilg, A.W. and Barr, S. (2005b), 'Encouraging Environmental Action by Exhortation: Evidence from a case study in Devon', *Journal of Environmental Planning and Management* 48:4, 593–618.

Gilg, A.W. and Barr, S. (2006), 'Behavioural Attitudes towards Water Saving: Evidence from a study of environmental actions', *Ecological Economics* 57:3, 400–14.

Global Action Plan (1999), *Small Change* (London: Global Action Plan).

Goldenhar, L.M. and Connell, C.M. (1992–93), 'Understanding and Predicating Recycling Behavior: An application of the theory of reasoned action', *Journal of Environmental Systems* 22:1, 91–103.

Golding, M. (1972), 'Obligations to Future Generations', *The Monist* 56, 85–99.

Goodwin, M. and Barr, S. (2007), 'The Politics of Scale in Climate Change Research: Enquiry, policy and activism', paper presented at the Association of American Geographers Conference in San Francisco, CA, April.

Gray, S. (1971), *The Electoral Register: Practical information for use when drawing samples, both for interview and postal survey* (London: OPCS).

Gregg, R. (1936), 'Voluntary Simplicity', *Visva Bharati Quarterly*, reprinted in *Manas* (4 September 1974).

Guagnano, G.A., Stern, P.C. and Dietz, T. (1995), 'Influences on Attitude-Behavior Relationships: A natural experiment with curbside recycling', *Environment and Behavior* 27, 699–718.

Guha, R. (2000), *Environmentalism: A global history* (Harlow: Longman).

Hajer, M.A. (2003), *Deliberative Policy Analysis: Governance in the network society* (Cambridge: Cambridge University Press).

Hallin, P.O. (1995), 'Environmental Concern and Environmental Behaviour in Foley, a Small Town in Minnesota', *Environment and Behavior* 27:4, 558–78.

Hamilton, L.C. (1983), 'Saving Water: A causal model of household conservation', *Sociological Perspectives* 26:4, 355–74.

Hardin, G. (1968), 'The Tragedy of the Commons', *Science* 162, 1243–48.

Harrison, C.M., Burgess, J. and Filius, P. (1996), 'Rationalizing Environmental Responsibilities – A Comparison of Lay Publics in the UK and the Netherlands', *Global Environmental Change* 6:3, 215–34.

Harvey, D. (1974), 'Population, Resources and the Ideology of Science', *Economic Geography* 50, 256–77.

Heberlein, T.A. and Warriner, G.K. (1983), 'The Influence of Price and Attitude on Shifting Residential Electricity Consumption from On- to Off-peak Periods', *Journal of Economic Psychology* 4, 107–30.

HM Treasury (2006), *Stern Review: The economics of climate change* (London: HM Treasury).

Heslop, L.A., Moran, L. and Cousineau, A. (1981), 'Consciousness in Energy Conservation Behavior: An exploratory study', *Journal of Consumer Research* 8, 299–305.

Hinchliffe, S. (1996), 'Helping the Earth Begins at Home. The Social Construction of Socio-Environmental Responsibilities', *Global Environmental Change* 6:1, 53–62.

Hines, J.M., Hungerford, H.R. and Tomera, A.N. (1987), 'Analysis and Synthesis of Research on responsible Environmental Behavior: A meta analysis', *Journal of Environmental Education* 18:2, 1–8.

Hobson, K. (2001), 'Sustainable Lifestyles: Rethinking barriers and behaviour change', in M.J. Cohen and J. Murphy (eds), *Exploring Sustainable Consumption: Environmental policy and the social sciences* (Amsterdam: Elsevier).

Hobson, K. (2002), 'Competing Discourses of Sustainable Consumption: Does the "rationalisation of lifestyles" make sense?', *Environmental Politics* 11, 95–120.

Hopper, J.R. and Nielsen, J.M. (1991), 'Recycling as Altruistic Behavior: Normative and behavioural strategies to expand participation in a community recycling programme', *Environment and Behavior* 23:2, 195–220.

Hutton, R.B. and McNeill, D.L. (1981), 'The Value of Incentives in Stimulating Energy Conservation', *Journal of Consumer Research* 8, 291–98.

Inglehart, R. (1990), *Culture Shift in Advanced Industrial Society* (Princeton, NJ: Princeton University Press).

Jackson, D.N. (1970), *Jackson Personality Inventory* (London, Ontario: University of Western Ontario).

Jackson, E.L. (1980), 'Perceptions of Energy Problems and the Adoption of Conservation Practices in Edmonton, Calgary', *Canadian Geographer* 24:2, 114–30.

Jackson, T. (2005), *Motivating Sustainable Consumption: a report to the Sustainable Development Research Network* (London: Sustainable Development Research Network).

Jacobs, M. (1991), *The Green Economy: Environment, sustainable development and the politics of the future* (London: Pluto Press).

Jordan, A. (ed.) *Environmental Policy in the European Union* (London: Earthscan).

Kaiser, F.G., Wolfing, S. and Fuher, U. (1999), 'Environmental Attitude and Ecological Behaviour', *Journal of Environmental Psychology* 19, 1–19.

Kantola, S.J., Syme, G.J. and Campbell, N.A. (1982), 'The Role of Individual Difference and External Variables in a Test of the Sufficiency of Fishbein's Model to Explain Behavioural Intentions to Conserve Water', *Journal of Applied Social Psychology* 12:1, 70–83.

Kantola, S.J., Syme, G.J. and Nesdale, A.R. (1983), 'The Effects of Appraised Severity and Efficacy in Promoting Water Conservation: An informational analysis', *Journal of Applied Social Psychology* 13, 164–82.

Karp, D.G. (1996), 'Values and their Effect on Pro-environmental Behavior', *Environment and Behavior* 28:1, 111–33.

Katzev, R.D. and Johnson, T.R. (1983), 'A Social-Psychological Analysis of Residential Electricity Consumption: The impact of minimal justification techniques', *Journal of Economic Psychology* 3, 267–84.

Kelly, M., Selman, P. and Gilg, A. W. (2004), 'Taking Sustainability Forward: Relating practice to policy in a changing legislative environment', *Town Planning Review* 75:3, 309–35.

Khan, M.A. (1995), 'Sustainable Development: The key concepts, issues ad implications', *Sustainable Development* 3:2, 63–9.

Kinnear, T.C., Taylor, J.R. and Ahmed, S.A. (1974), 'Ecologically Concerned Consumers: Who are they?', *Journal of Marketing* 38, 20–24.

Kok, G. and Siero, S. (1985), 'Tin Recycling: Awareness, comprehension, attitude, intention and behavior', *Journal of Economic Psychology* 6, 157–73.

Lam, S. (1999), 'Predicting Intentions to Conserve Water from the Theory of Planned Behaviour, Perceived Moral Obligation and Perceived Water Right', *Journal of Applied Social Psychology* 29, 1058–71.

LaPiere, R.T. (1934), 'Attitudes vs. Actions', *Social Forces* 13:2, 230–37.

Lee, J.A. and Holden, S.J.S. (1999), 'Understanding the Determinants of Environmentally Conscious Behavior', *Psychology and Marketing* 16:5, 373–92.

Lele, S.M. (1991), 'Sustainable Development – A Critical Review', *World Development* 19:6, 607–21.

Leonard-Barton, D. (1981), 'Voluntary Simplicity Lifestyles and Energy Conservation', *Journal of Consumer Research* 8, 243–52.

Linnros, H.D. and Hallin, P.O. (2001), 'The Discursive Nature of Environmental Conflicts: The case of the Öresund link', *Area* 3: 4, 391–403.

Littlewood, S. and While, A. (1997), 'A New Agenda for Governance? Agenda 21 and the Prospects for Holistic Decision Making', *Local Government Studies* 23:4, 110–23.

Local Government Management Board (LGMB) (1994), *Local Agenda 21: Principles and Process: A step by step guide* (Luton: LGMB).

Luyben, P.D. (1982), 'Prompting Thermostat Setting Behavior: Public response to a Presidential appeal for conservation', *Environment and Behavior* 14:1, 113–28.

Macey, S.M. and Brown, M.A. (1983), 'Residential Energy Conservation: The role of past experience in repetitive household behavior', *Environment and Behavior* 15:2, 123–41.

Mackenzie, D. (1990), 'The Green Consumer', *Food Policy* 15:6, 461–66.

Macnaghten, P. and Jacobs, M. (1997), 'Public Identification with Sustainable Development: Investigating cultural barriers to participation', *Global Environmental Change* 7:1, 5–24.

MacNaghten, P. and Urry, J. (1998), *Contested Natures* (London: Sage).

Mainieri, T., Barnett, E.G., Valdero, T.R., Unipan, J.B. and Oskamp, S. (1997), 'Green Buying: The influence of environmental concern on consumer behavior', *Journal of Social Psychology* 137:2, 189–204.

Malthus, T. (1798), *An Essay on the Principle of Population* (London: Macmillan).

Manning, E.W. (1990), 'Sustainable Development: The challenge', *Canadian Geographer* 34:4, 290–302.

Maslow, A.H. (1970), *Motivation and Personality* (London: Harper and Row).

McCormick, J. (1989), *The Global Environmental Movement*, 2nd edn (London: Belhaven).

McDougall, G.H.G., Claxton, J.D., Ritchie, J.R.B. and Anderson, C.D. (1981), 'Consumer Energy Research: A review', *Journal of Consumer Research* 8, 343–54.

Meacher, M. (1999), 'Sustainable Development: A better quality of life', *Eg Magazine* 5:6, 2–3.

Meacher, M. (2002), 'Putting Theory into Practice: Towards action at the local and community level' *Eg Magazine* 8:10, 4–5.

Meadowcroft, J. (1999), 'The Politics of Sustainable Development: Emergent arenas and challenges for political science', *International Political Science Review* 20:2, 219–37.

Meadows D.H., Meadows, D.L. and Randers, D. (1992), *Beyond the Limits: Global collapse or a sustainable future* (London: Earthscan).

Meadows, D.H., Randers, J. and Berhens III, W. (1972), *The Limits to Growth* (New York: Universe Books).

Midden, G.J.H. and Ritsema, B.S.M. (1983), 'The Meaning of Normative Processes for Energy Conservation', *Journal of Economic Psychology* 4, 37–55.

Milton, K. (1993), *Environmentalism: The view from anthropology* (London: Routledge).

Minton, A.P. and Rose, R.L. (1997), 'The Effect of Environmental Concern on Environmentally Friendly Consumer Behavior: An exploratory study', *Journal of Business Research* 40, 37–48.

Moffatt, I. (1996), *Sustainable Development: Principles, analysis and policies* (London: Parthenon).

Moore, S., Murphy, M. and Watson, R. (1994), 'A Longitudinal Study of domestic Water Conservation Behaviour', *Population and Environment* 16:2, 175–89.

Morris, J. (1999), 'Chasing the Millennium Deadline: Is Local Agenda 21 on target?', *Eg Magazine* 5:4, 2–5.

Munton, R. (1997), 'Engaging Sustainable Development: Some observations on progress in the UK', *Progress in Human Geography* 21:2, 151–9.

Munton, R. and Collins, K. (1998), 'Government Strategies for Sustainable Development', *Geography* 83, 356–7.

Myers, G. and Macnaghten, P. (1998), 'Rhetorics of Environmental Sustainability: Commonplaces and places', *Environment and Planning A* 30:2 333–53.

Myers, N. (1987), 'The Environmental Basis of Sustainable Development', *Annals of Regional Science* 21:3, 33–43.

Nancarrow, B.E., Smith, L.M. and Syme, G.J. (1996–97), 'The Ways People Think about Water', *Journal of Environmental Systems* 25:1, 15–27.

National Consumer Council (NCC) and The New Economics Foundation (NEF) (2004), *Carrots not Sticks: The possibilities of a sustainable consumption reward card for the UK* (London: NCC).

Noe, F.P. and Snow, R. (1990), 'The New Environmental Paradigm and Further Scale Analysis', *Journal of Environmental Education* 21, 20–26.

Norgaard, R.B. (1988), 'Sustainable Development: A co-evolutionary view', *Futures* 20:6, 606–20.

Office of the Deputy Prime Minister (ODPM) (2005), *Planning Policy Statement 1: Delivering sustainable development* (London: ODPM).

Olli, E., Grendstad, D. and Wollebark, D. (2001), 'Correlates of Environmental Behaviors: Bringing back social context', *Environment and Behavior* 33:2, 181–208.

Olsen, M.E. (1981), 'Consumers' Attitudes toward Energy Conservation', *Journal of Social Issues* 37:2, 108–31.

O'Riordan, T. (1976), *Environmentalism* (London: Pion).

O'Riordan, T. (1985), 'Future Directions in Environmental Policy', *Environment and Planning A* 17, 1431–46.

O'Riordan, T. (1989), 'The Challenge for Environmentalism', in R. Peet and N. Thrift (eds), *New Models in Geography* (London: Unwin Hyman).

O'Riordan, T. (1993), 'The Politics of Sustainability', in R.K. Turner (ed.), *Sustainable Environmental Economics and Management* (London: Belhaven), 37–69.

O'Riordan, T. (1997a), 'Sustainability and New Labour Radicalism', *ECOS* 18:1, 12–15.

O'Riordan, T. (1997b), 'Labour's Greenish Credentials', *ECOS* 18:2, 2–5.

Oskamp, S. (2000a), 'A Sustainable Future for Humanity? How Can Psychology Help?', *American Psychologist* 55:5, 496–508.

Oskamp, S. (2000b), 'Psychology of Promoting Environmentalism: Psychological contributions to achieving an ecologically sustainable future for humanity', *Journal of Social Issues* 56:3, 373–90.

Oskamp, S., Harrington, M.J., Edwards, T.C., Sherwood, D.L., Okuda, S.M. and Swanson, D.C. (1991), 'Factors Influencing Household Recycling Behavior', *Environment and Behavior* 23:4, 494–519.

Owens, S. (2000), 'Engaging the Public: Information and deliberation in environmental policy', *Environment and Planning A* 32, 1141–48.

Paehlke R.C. (1989), *Environmentalism and the Future of Progressive Politics* (London: Yale University Press).

Painter, J., Semenik, R. and Belk, R. (1983), 'Is There a Generalized Conservation Ethic? A Comparison of the Determinants of Gasoline and Home Heating Energy Conservation', *Journal of Economic Psychology* 3, 317–31.

Pearce, D. (1988), 'Economics, Equity and Sustainable Development', *Futures* 20:6, 598–605.

Pearce, D. (1991), *Blueprint 2: Greening the world economy* (London: Earthscan in association with the London Environmental Economics Centre).

Pearce, D. (1993), *Blueprint 3: Measuring sustainable development* (London: Earthscan).

Pearce, D., Markandya, A. and Barbier, E. (1989), *Blueprint for a Green Economy* (London: Earthscan).

Pellow, D.N. (2003), *Garbage Wars: The struggle for environmental justice in Chicago* (Cambridge, MA: MIT Press).

Pepper, D. (1986), *The Roots of Modern Environmentalism* (London: Routledge).

Pepper, D. (1993), *Eco-socialism from Deep Ecology to Social Justice* (London: Routledge).

Pepper, D. (2003), *Modern Environmentalism* (London: Routledge).

Petts, J. (1995), 'Waste Management Strategy Development: A case study of community involvement and consensus-building in Hampshire', *Journal of Environmental Planning and Management* 38:4, 519–36.

Pezzy, J. (1992), 'Sustainability: An inter-disciplinary guide', *Environmental Values* 1, 321–62.

Porritt, J. and Levett, R. (1999), 'A Better Quality of Strategy? Assessing the White Paper on Sustainable Development', *Eg Magazine* 5:6, 4–7.

Redclift, M. (1988), 'Sustainable Development and the Market: A framework for analysis', *Futures* 20:6, 635–50.

Redclift, M. (1991), 'The Multiple Dimensions of Sustainable Development', *Geography* 76, 36–42.

Rees, J. (1991), 'Equity and Environmental Policy', *Geography* 76, 292–303.

Riddell, R. (1981), *Ecodevelopment: Economics, ecology and development an alternative to growth imperative models* (Farnborough: Gower).

Ritchie, J.R.B., McGougall, G.H.G. and Claxton, J.D. (1981), 'Complexies of Household Energy Consumption and Conservation', *Journal of Consumer Research* 8, 233–42.

Roberts, J.A. (1993), 'Sex Differences in Socially Responsible Consumers' Behavior', *Psychological Reports* 73, 139–48.

Roberts, J.A. (1996), 'Green Consumers in the 1990s: Profile and implications for advertising', *Journal of Business Research* 36, 217–31.

Roberts, J.A. (2004), *Environmental Policy* (London: Routledge).

Roberts, J.A. and Bacon, D.R. (1997), 'Exploring the Subtle Relationships between Environmental Concern and Ecologically Conscious Consumer Behavior', *Journal of Business Research* 40, 79–89.

Rose, N. and Miller, P. (1992), 'Political Power and the State: Problematics of government', *British Journal of Sociology* 43:2, 173–205.

Ross, A. (2000), 'LA21 Evaluation: Are there lessons for community strategies?', *Eg Magazine* 6:10, 6–8.

Ross, A. (2005), 'The UK Approach to Delivering Sustainable Development in Government: A case study in joined-up working', *Journal of Environmental Law* 17:1, 27–49.

Sadalla, E.K. and Krull, J.L. (1995), 'Self-presentational Barriers to Resource Conservation', *Environment and Behavior* 27:3, 328–53.

Samuelson, C.D. and Biek, M. (1991), 'Attitudes toward Energy Conservation: A confirmatory factor analysis', *Journal of Applied Social Psychology* 21:7, 549–68.

Schahn, J. and Holzer, E. (1990), 'Studies of Individual Environmental Concern: The role of knowledge, gender and background variables', *Environment and Behavior* 22:6, 767–86.

Schuhwerk, M.E. and Lefkoff-Hagius, R. (1995), 'Green or Non-green? Does type of Appeal Matter when Advertising a Green Product', *Journal of Advertising* 24:2, 45–54.

Schultz, P.W., Oskamp, S. and Mainieri, T. (1995), 'Who Recycles and When? A Review of Personal and Situational Factors', *Journal of Environmental Psychology* 15, 105–21.

Schwartz, S.H. (1977), 'Normative Influences on Altruism', in L. Berkowitz (ed.), *Advances in Experimental Social Psychology* 10, 221–79.

Schwartz, S.H. (1992), 'Universals in the Content and Structure of Values: Theoretical advances and empirical test in 20 countries', *Advances in Experimental Social Psychology* 25, 1–65.

Schwartz, S.H. and Blisky, W. (1987), 'Toward a Psychological Structure of Human Values', *Journal of Personality and Social Psychology* 53, 550–62.

Schwepker, C.H. and Cornwell, T.B. (1991), 'An Examination of Ecologically Concerned Consumers and their Intention to Purchase Ecologically Packaged Products', *Journal of Public Policy and Marketing* 10:2, 77–101.

Scott, D. and Willits, F.K. (1994), 'Environmental Attitudes and Behavior: A Pennsylvania survey', *Environment and Behavior* 26:2, 239–60.

Segun, C., Pelletier, L.G. and Hunsley, J. (1998), 'Towards a Model of Environmental Activism', *Environment and Behavior* 30, 628–52.

Seligman, C., Kriss, M., Darley, J.M., Fazio, R.H., Becker, L.J. and Payor, J.B. (1979), 'Predicting Summer Energy Conservation from Homeowners' Attitudes', *Journal of Applied Social Psychology* 9:1, 70–90.

Seligman, C., Syme, G.J. and Gilchrist, R. (1994), 'The Role of Values and Ethical Principles in Judgements of Environmental Dilemmas', *Journal of Social Issues* 50:3, 105–19.

Selman, P. (1996), *Local Sustainability: Planning and managing ecologically sound places* (London: Chapman).

Selman, P. (1998), 'Local Agenda 21: Substance or spin?', *Journal of Environmental Planning and Management* 41:5, 533–53.

Selman, P. and Parker, J. (1997), 'Citizenship, Civicness and Social Capital in Local Agenda 21', *Local Environment* 2:2, 171–84.

Selman, P. and Parker, J. (1999), 'Tales of Local Sustainability', *Local Environment* 4:1, 47–60.

Seyfang, G. (2003), *From Frankenstein Foods to Veggie Box Schemes: Sustainable consumption in cultural perspective* (Norwich: CSERGE Working Paper EDM 03–13, CSERGE).

Shrum, L.J., McCarty, J.A. and Lowrey, T.M. (1995), 'Buyer Characteristics of the Green Consumer and their Implications for Advertising Strategy', *Journal of Advertising* 24:2, 71–82.

Simmons, I. (1995), 'Green Geography: An evolving recipe', *Geography* 80, 139–45.

South West Regional Assembly (2004), *Just Connect! A Regional Strategy for South West England 2004–2016* (Taunton: SWRA).

Sparks, P. and Shepherd, R. (1992), 'Self-identity and the Theory of Planned Behavior: The role of identification with "green consumerism"', *Social Psychology Quarterly* 55:4, 388–99.

Stead, W.E., Stead, J.G. and Worrell, D.L. (1991), 'Investigating the Psychology of the Green Consumer', *Psychological Reports* 68, 833–34.

Stern P.C. (1992a), 'What Psychology Knows about Energy Conservation', *American Psychologist* 47, 1224–32.

Stern, P. C. (1992b), 'Psychological Dimensions of Global Environmental Change', *Annual Review of Psychology* 43, 269–309.

Stern, P.C. (2000), 'Psychology and the Science of Human-Environment Interactions', *American Psychologist* 55:5, 523–30.

Stern, P.C. and Oskamp, S. (1987), 'Managing Scarce Environmental Resources', in D. Stokols and I. Altman (eds), *Handbook of Environmental Psychology* Vol. 2 (New York: Wiley), 1043–88.

Stern, P., Dietz, T. and Guagnano, G.A. (1995), 'The New Ecological Paradigm in Social-psychological Context', *Environment and Behavior* 27:6, 723–43.

Stone, G., Barnes, J.H. and Montgomery, C. (1995), 'ECOSCALE: A scale for the measurement of environmentally responsible consumers', *Psychology and Marketing* 12:7, 595–612.

Strong, M.F. (1977), 'The International Community and the Environment', *Environmental Conservation* 4:3, 165–72.

Sustainable Development Commission (2004), *Shows Promise, But Must Try Harder* (London: Sustainable Development Commission).

Sustainable Development Commission (2005), 'Better, Bolder, Braver? The Sustainable Development Commission comments on the government's new strategy commitments', *Eg Magazine* 11 (2–3), 1–3.

Syme, G.J.M., Nancarrow, B.E. and Seligman, C. (2000), 'The Evaluation of Information Campaigns to Promote Voluntary Household Water Conservation', *Evaluation Review* 24: 6, 539–78.

Syme, G.J.M., Seligman, C. and Thomas, J.F. (1990–1991), 'Predicting Water Consumption from Homeowners' Attitudes', *Journal of Environmental Systems* 20:2, 157–68.

Taylor, S. and Todd, P. (1997), 'Understanding the Determinants of Consumer Composting Behaviour', *Journal of Applied Social Psychology* 27, 602–28.

Thompson, S.C.G. and Barton, M.A. (1994), 'Ecocentric and Anthropocentric Attitudes toward the Environment', *Journal of Environmental Psychology* 14, 149–57.

Thorson, E., Page, T. and Moore, J. (1995), 'Consumer Response to four Categories of "Green" Television Commercials', *Advances in Consumer Research* 22, 243–50.

Tucker, L.R. Jr (1980), 'Identifying the Environmentally Responsible Consumer: The role of internal-external control of reinforcements', *Journal of Consumer Affairs* 14:2, 326–40.

Turner, R.K. (1993), *Sustainable Environmental Economics and Management: Principles and practice* (London: Belhaven).

Tuxworth, B. (1996), 'From Environment to Sustainability: Surveys and analysis of Local Agenda 21 process development in UK local authorities', *Local Environment* 1:3, 277–97.

United Nations Conference on Environment and Development (UNCED) (1992), *Agenda 21 – Action Plan for the Next Century*, endorsed at UNCED (New York: UNCED).

Van Houwelingen, J.H. and Van Raaij, W.F. (1989), 'The Effect of Goal Setting and Daily Electronic Feedback on In-home Energy Use', *Journal of Consumer Research* 16, 98–105.

Van Liere, K. and Dnlap, R.E. (1980), 'The Social Bases of Environmental Concern: A review of hypotheses, explanations and empirical evidence', *Public Opinion Quarterly* 44:2, 181–97.

Van Raaij, W.F. and Verhallen, T.M.M. (1983), 'A Behavioral Model of Residential Energy Use', *Journal of Economic Psychology* 3, 39–63.

Verhallen, T.M.M. and Van Raaij, W.F. (1981), 'Household Behavior and the Use of Natural Gas for Home Heating', *Journal of Consumer Research* 8, 253–7.

Vining, J. and Ebreo, A. (1990), 'What Makes a recycler? A Comparison of Recyclers and Nonrecyclers', *Environment and Behavior* 22:1, 55–73.

Vining, J. and Ebreo, A. (1992), 'Predicting Recycling Behaviour from Global and Specific Environmental Attitudes and Changes in Recycling Opportunities', *Journal of Applied Social Psychology* 22, 1580–607.

Waks, L.J. (1996), 'Environmental Claims and Citizen Rights', *Environmental Ethics* 18:2, 133–48.

Warriner, G.K., McDougall, G.H.G. and Claxton, J.D. (1984), 'Any Data or None At All? Living with Inconsistencies in Self-reports of Residential Energy Consumption', *Environment and Behavior* 16:4, 503–26.

Webster, F.E. Jr (1975), 'Determining the Characteristics of the Socially Conscious Consumer', *Journal of Consumer Research* 2, 188–96.

Weigel, R. and Weigel, J. (1978), 'Environmental Concern: The development of a measure', *Environment and Behavior* 10, 3–15.

Weinberg, R.S., Pellow, D.N. and Schnaiberg, A. (2001), *Urban Recycling and the Search for Sustainable Community Development* (Princeton, NJ: Princeton University Press).

White, L. (1967), 'The Roots of our Ecological Crisis', *Science* 155, 203–7.

Widegren, O. (1998), 'The New Environmental Paradigm and Personal Norms', *Environment and Behavior* 30:1 75–100.

Wilbanks, T.J. (1994), '"Sustainable Development" in Geographic Perspective', *Association of American Geographers* 84:4, 541–56.

Williams, C.C. and Millington, A.C. (2004), 'The Diverse and Contested Meanings of Sustainable Development', *The Geographical Journal* 170:2, 99–104.

Wilson, G. (1997), 'Farmer Environmental Attitudes and ESA Participation', *Geoforum* 27, 115–31.

Wilson, G. (2000), 'Putting Sustainable Development at the Heart of Government Decision Making – an Update', *Eg Magazine* 6:3, 13–14.

Wilson, G. (2001), 'From Productivism to Post-productivism ... and Back Again? Exploring the (Un)changed Natural and Mental Landscapes of European Agriculture', *Transactions of the Institute of British Geographers* 26:1, 77–102.

Winett, R.A. and Ester, P. (1983), 'Behavioral Science and Energy Conservation: Conceptualizations, strategies, outcomes, energy policy applications', *Journal of Economic Psychology* 3, 203–29.

Winett, R.A. (1987), 'Comment on Costanzo et al.'s Energy Conservation Behavior: The difficult path from information to action', *American Psychologist* 42, 957–8.

World Commission on Environment and Development (WCED) (1987), *Our Common Future* (Oxford: Oxford University Press).

Young, J. (1990), *Post Environmentalism* (London: Belhaven).

Zinkhan, J.M. and Carlson, L. (1995), 'Green Advertising and the Reluctant Consumer', *Journal of Advertising* 24:2, 1–6.

Index

Agenda 21 28–9, 32, 45, 48, 52–4, 56, 71

Behaviour change
  civic approach 18, 86, 92–3, 97, 254, 258
  barriers 67–8, 82, 93, 97, 101–3, 119,
    125, 131–3, 143, 162, 185, 187–8,
    191–2, 225, 229–30, 233, 236, 238,
    240, 242, 244, 246–7, 258–9
  deliberative approach 18, 47, 86, 91–3,
    96–7, 254, 258
  information-intensive approach 97
  linear model 82, 93, 97, 101, 117, 130,
    186, 191, 224, 253–4, 258
  rationalistic approach 18, 86, 96–7, 101,
    105, 191, 254
behavioural commitment 87, 102, 114–16,
    118–19, 121, 124, 131, 134–5, 138,
    170, 186–8, 191, 209, 232, 240, 243,
    253, 259
Brundtland Report 17, 24, 26–8, 31–2, 34,
    38, 43, 45, 47–8, 55, 57–8, 69, 251–2

Carson, R., *Silent Spring* 9–10
campaigns
  Are You Doing Your Bit? 86, 89–90, 102,
    107, 129
  Going for Green 86–9, 129, 253
  Helping the Earth Begins at Home 86,
    88, 96, 129
Club of Rome 13–14, 16, 22, 25
cluster analysis 159, 163–4, 171
community strategies 13–14, 16, 22, 25, 49,
    53–4, 73–4, 76, 82, 91, 135

Darnton, A. 90, 162–3, 191, 253
decision-making 45, 47, 54, 59, 80, 82,
    135, 144, 252, 254
De Young, R. 120–21, 143

Eden, S. 5, 36, 92–6, 254
environmental
  awareness 30, 118, 239
  commitment 9, 103, 159, 155

'crisis' 3, 11, 17, 22, 25, 47
  policy 8, 21, 26, 29, 52, 54–5, 59, 86,
    90–91, 93–4, 114, 252
  values 33–4, 101–2, 111–13, 115, 124,
    130–31, 138–40, 163–4, 176, 178–9,
    187, 193–4, 209, 215, 222, 255, 257
environmental behaviour(s)
  compartmentalised 88, 102, 117, 125,
    129, 222
  energy saving 6, 35, 100, 106–8, 119–20,
    122, 132, 134–6, 138, 142, 144–5,
    148, 155, 158, 222
  'green consumption' 106–7, 109–11,
    17, 122–3, 129, 132, 134–8, 148–9,
    155–6, 158, 183–4, 202, 209, 222–3
  recycling 3, 37, 51, 56, 81, 87, 99–100,
    105–7, 109, 114–17, 119–21, 129,
    131–2, 135–6, 139, 143, 150, 155–8,
    163, 165–71, 183–4, 192, 195–7,
    200–203, 214, 216–23, 230–34,
    240–41, 244, 249, 256
  waste management 19, 53, 75, 95, 106,
    109, 111, 115, 117, 131, 136–8,
    150–51, 156–7, 165, 187, 232
  water conservation 106–8, 118, 132,
    134–8, 146–8, 155–8, 170, 183, 232,
    243, 249
environmental psychology 25, 86, 99,
    105–6, 116, 118, 255
Environmental Protection Act (1995) 59
environmentalism 9, 44, 47–8, 115, 175–6,
    186, 246
European Union (EU) 48, 52–5, 150

factor analysis 152–3, 155, 159, 176, 179,
    184

Gibbs, D. 44, 77–78
Gilg, A.W. 3, 45, 71, 80, 86, 111, 149–50,
    164, 254

Hardin, G., *Tragedy of the Commons* 6–7,
    23

indicators of sustainability 61–5, 69–70, 79, 114

lifestyle groups 18–20, 82, 103, 112, 119, 125, 130–32, 144, 148, 158–9, 161–4, 171, 176, 179, 185, 187, 1189, 192, 202, 222–4, 229, 231, 233, 240, 242–9, 255–60
'Limits to Growth' 4, 11–12, 14–15, 22, 26, 179, 251
Local Agenda 21 48, 74–81,135

marketing
  branding 103, 143, 162, 233, 236–8, 246–8, 258–9, 261
  social 19, 107, 162–3, 225, 229–31, 233, 235–7, 239, 241–3, 245–9, 258–9

natural capital 34, 37–8, 44, 61, 66, 69
new communitarianism 60, 80, 123

O'Riordan, T. 16, 33, 59, 114

path analysis 159, 194–5, 198–200, 202, 205–8, 211–14, 217–20
Pearce, D. 35, 37, 44–46, 60, 77, 114
planning 40, 49, 52, 59, 71, 73, 75–7, 91, 95
psychological factors 19, 119, 163, 182, 184–5, 193–4, 201, 209–10, 215–16, 221–3, 234, 255, 257

qualitative 77, 91, 125, 131–2, 159, 195, 225, 231, 241–2, 244, 260
quantitative 22–3, 25, 78, 97–8, 105, 125, 132, 1580–9, 162, 192, 195, 225, 230, 233, 235, 238, 241–2, 244, 254, 257, 260

Rio Earth Summit 17, 29, 51–3, 74–5, 82

Roberts, J.A. 44, 110, 117, 122–3, 136, 183–4, 222

Schwartz, S.H. 33, 95, 99, 112–13, 123, 139, 176
Securing the Future 66–8
segmentation 19, 103, 125, 161–3, 165, 171, 189–90, 202, 230, 242, 257–9
Selman, P. 48, 74, 78, 80–81, 123
situational characteristics 19, 102, 146, 163, 194, 201, 223, 255
social-psychological 86, 97–8, 101, 105, 111, 132, 191–2, 224, 254–5, 258
  framework approach 99, 101
  modelling approach 99
Stern, P.C. 107–8, 113, 116–17, 139, 176, 223
Stern Review 3
sustainable consumption 65–6, 85, 89, 245
sustainable development models
  harmonisation 31–2, 41–2
  nested 31–2, 42–3
  three rings 41–3
  weak/strong 17, 32, 44–6, 57, 69, 77, 88
Sustainable Development Commission 47–8, 63–4
Sustainable Development Strategy 48, 51–3, 58, 60, 64–5, 76, 82, 85–7, 89–90, 101, 191, 252

technocentric 33, 35, 43, 46, 48, 56, 69, 86, 112, 114–15, 139, 179, 188, 251
  ecological modernisation 29, 35, 69, 81
Theory of Planned Behaviour 100–101, 122, 184, 192, 255
This Common Inheritance 57
trust, (mis-) 78, 92–4, 96, 123, 181, 183, 185–7, 221, 237, 240, 244

United Nations 17, 23–4, 26–30, 53, 251–2